T0214196

Modern Birkhäuser Classics

Many of the original research and survey monographs in pure and applied mathematics, as well as textbooks, published by Birkhäuser in recent decades have been groundbreaking and have come to be regarded as foundational to the subject. Through the MBC Series, a select number of these modern classics, entirely uncorrected, are being re-released in paperback (and as eBooks) to ensure that these treasures remain accessible to new generations of students, scholars, and researchers.

Robert Roussarie

Bifurcations of Planar Vector Fields and Hilbert's Sixteenth Problem

Reprint of the 1998 Edition

 Birkhäuser

Robert Roussarie
Institut de Mathématique
Université de Bourgogne
Dijon, France

ISBN 978-3-0348-0717-3   ISBN 978-3-0348-0718-0   (eBook)
DOI 10.1007/978-3-0348-0718-0
Springer Basel Heidelberg New York Dordrecht London

Library of Congress Control Number: 2013953592

Mathematics Subject Classification (2010): 58-XX, 34C05, 34C23, 35B32, 37G15, 37G99, 37-02

© Springer Basel 1998
Reprint of the 1st edition 1998 by Birkhäuser Verlag, Switzerland
Originally published as volume 164 in the Progress in Mathematics series

Cover design: deblik, Berlin

Printed on acid-free paper

Springer Basel is part of Springer Science+Business Media
(www.birkhauser-science.com)

Contents

Preface

Let $X(x,y) = P(x,y)\, \dfrac{\partial}{\partial x} + Q(x,y)\, \dfrac{\partial}{\partial y}$ be a differentiable planar vector field. By integrating the differential equation $\dot{x} = P(x,y)$, $\dot{y} = Q(x,y)$, one obtains *the flow* $\varphi(t;(x,y))$ of X (the trajectories with initial conditions $(x,y) \in \mathbb{R}^2$).

In spite of the restriction to dimension 2, flows of planar vector fields model many interesting phenomena, for example the Van der Pol equation: $\ddot{x} + \alpha(x^2 - 1)\dot{x} + x = 0$, which models an electric circuit including a triode valve; or the Duffin equation: $\ddot{x} + k\dot{x} - x + x^3 = 0$, which models a buckled beam (see [GH]). These second order differential equations are equivalent to differential equations associated to vector fields defined on $\mathbb{S}^1 \times \mathbb{R}$. This last set can be identified with an open subset of the plane. Of course, these equations do not include any exterior forcing. If we add a forcing periodic term to the equation (for instance $\varepsilon \cos \Omega t$), the system would be equivalent to an equation of a vector field in $\mathbb{S}^1 \times \mathbb{R}^2$, i.e., in a three-dimensional space, with a much more complicated dynamic than in the plane (it may have some chaotic orbits). Nevertheless, even in this case the study of the unforced equation will be an unavoidable first step in the study of the weakly forced equation.

The flow of a planar vector field X is a very simple example of a dynamical system. In fact, by the Poincaré-Bendixson theorem, every recurrent orbit of the flow φ is trivial, i.e., a singular point or a periodic orbit. Hence, it is quite easy to describe *the phase portrait* of the vector field X once one knows these trivial orbits. Recall that the phase portrait of X is the equivalence class of X up to the action of homeomorphisms sending oriented orbits to oriented orbits.

Let us consider for instance, the equation of a simple pendulum with a linear dissipation,

$$\ddot{x} + k\dot{x} + \omega^2 \sin x = 0, \text{ with } k > 0.$$

In the phase space $\mathbb{S}^1 \times \mathbb{R}$ with coordinates $x \in \mathbb{S}^1$ and $y = \dot{x} \in \mathbb{R}$, this equation is equivalent to the equation of the vector field,

$$X = y\frac{\partial}{\partial x} - (ky + \omega^2 \sin x)\frac{\partial}{\partial y}.$$

As a consequence of the Poincaré-Bendixson theorem, the phase portrait of X is uniquely determined by the following observations. There exist only two singular points: a stable focus at $(0,0)$ and a saddle point at $(\pi,0)$. Next, there exist no periodic orbits, because the divergence of X is equal to k. Finally, the ω-limit set of any orbit must be a singular point. The phase portrait determined by these observations is represented in Figure 1.

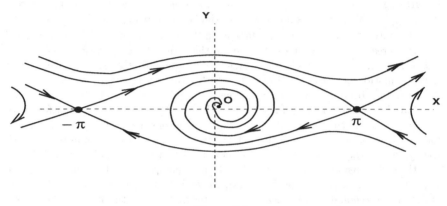

Figure 0.1

Hence, in order to know the dynamics of the flow, it is crucial to determine the position and the nature of its *singular elements* (the singular points and the periodic orbits). The difficulty of this task depends heavily on the way the vector field is defined. For instance, if the vector field is polynomial, i.e., if the components P, Q of the vector field are polynomials, the determination of the singular points is an algebraic question. Moreover, one can obtain the phase portrait of X near each singular point, at least when this point is algebraically isolated, using Seidenberg's theorem which can be translated into an algebraic algorithm.

On the other hand, even in the simple case of polynomial vector fields, it is, in general, impossible to determine explicitly, i.e., with an algebraic algorithm, the position and the nature of periodic orbits. Roughly speaking, the reason is the following: periodic orbits are obtained by equations in terms of flow, which cannot in general be computed explicitly. As a consequence, even if the dynamic generated by the flow of a planar vector field is very simple, its explicit determination is not possible in general. To study it, one has to use the qualitative ideas first introduced by Poincaré [Poi]. In some cases, they enable a study of the flow, without having to integrate the differential equation.

In this text, we study *families of planar vector fields* (X_λ), depending on a parameter $\lambda \in \mathbb{R}^k$. The phase space will be $\mathbb{R}^2$ or, more generally, a surface of genus 0. Parameters enter naturally into the description of the system which is modeled by a differential equation (the temperature of ambient space, the length

of the pendulum which determines the oscillation ω in the above example, etc.
...). However, even if one is only interested in the study of vector fields, it may be
interesting to consider vector fields which depend on parameters. For instance, it
will be convenient to consider the polynomial vector fields of degree $\leq n$ as a family
X_λ^n, whose parameter λ belongs to the set of coefficients of the two components
P, Q.

Let us consider a family (X_λ). For each value of λ, one can consider the
phase portrait of X_λ. As we have suggested above, if X_λ has only a finite number of singular elements, the phase portrait of X_λ on each compact region has a
combinatorial description. However, in order to understand the family, it is not
sufficient to understand the phase portrait of each X_λ. This would be useless. We
need the less static approach of *bifurcation theory*. This is in the study of how the
phase portrait changes (bifurcates) when the parameter λ varies in the parameter
space. Let us give a rough idea of this problem. For a given family (X_λ), there
is a closed subset $\Sigma \subset \mathbb{R}^k$, the parameter space, for which the phase portrait of
X_λ changes. In general, Σ is a stratified set and the degeneracy of X_λ increases
with the codimension of the strata to which λ belongs. The open set $U = \mathbb{R}^k \backslash \Sigma$ is
the set where X_λ is structurally stable (this means that the phase portrait of X_λ
does not change for small perturbations of λ). If $\lambda \in U$, and restricting ourselves
to a compact domain in the phase space, X_λ has only a finite number of singular
elements: singular points and isolated periodic orbits, which are also called *limit
cycles*. When λ crosses Σ one may have bifurcations which imply a change in the
number of some singular elements.

The difference that we have pointed out between the study of singular points
and the study of periodic orbits for an individual vector field is even more important in the study of their bifurcations in families: if we consider a polynomial
family, the bifurcations among singular points are again given by algebraic algorithms and this will not be the case, in general, for the limit cycles.

In order to study bifurcations of limit cycles we have first to detect where
in the phase space such bifurcations can occur. Of course, it is there where limit
cycles accumulate. Hence, it is natural to introduce these sets of accumulation. We
call them *limit periodic sets* of the family. More precisely, a limit periodic set Γ,
for the value λ_* of the parameter, will be any limit of a sequence of limit cycles
$(\gamma_n)_n$. The limit of this sequence is taken in the Hausdorff topology of the set of
all compact subsets of $\mathbb{R}^2$. We suppose that each γ_n is a limit cycle for a vector
field X_{λ_n}, and that the sequence $(\lambda_n)_n$ tends to some λ_* in the parameter space.
To understand the behaviour of the limit cycles in the family (X_λ), one has to
understand how they bifurcate from each periodic limit set Γ of the family.

*The study of bifurcations of limit periodic sets in families of planar
vector fields is precisely the subject of this book.*

In this bifurcation problem, the first step is to obtain a bound for the number
of limit cycles which bifurcate from Γ. Such a bound is called the *cyclicity* of Γ

in (X_λ) and is denoted by $\mathcal{C}ycl$ (X_λ, Γ). Of course, this bound depends only on the germ of the family along Γ, also called the *unfolding* (X_λ, Γ). It is easy to find differentiable unfoldings of limit periodic sets with infinite cyclicity. Hence, in order to be able to prove that the cyclicity is finite, we have to restrict ourselves to typical unfoldings. A general conjecture is that the cyclicity is finite for any *analytic* unfolding. It will be proved in the text that this conjecture would imply a positive answer to the "existential" Hilbert's 16th problem:

> *"Prove that for any $n \geq 2$, there exists a finite number $H(n)$ such that any polynomial vector field of degree $\leq n$ has less that $H(n)$ limit cycles".*

Besides general questions, as is the above conjecture or Hilbert's problem, it is important to be able to compute the cyclicity for explicit typical unfoldings (X_λ, Γ). A famous example of such an explicit computation is the result of Bautin [B], proving that each focus or center singular point in the quadratic polynomial family X_λ^2 has a cyclicity of at most three.

Once one has detected the periodic limit sets of a family and computed their cyclicity, it remains to study their local bifurcations in the family. In general, it is then easy to finish the study of the *bifurcation diagram*, i.e., to obtain the set Σ and to describe the bifurcations occurring by crossing this set. To illustrate these general considerations, we want to introduce now a simple example of a family: the *Bogdanov-Takens unfolding*. This example only depends on two parameters and the diagram of bifurcation is easy to describe, but the study is sufficiently nontrivial to give a good idea of the general problem. Moreover, this example will be used throughout the text to illustrate the different methods we want to introduce.

The Bogdanov-Takens unfolding is a local 2-parameter family near a singular point of a vector field X_0 whose 2-jet is equivalent to:

$$X_0^{N\pm} \begin{cases} \dot{x} = y \\ \dot{y} = x^2 \pm xy. \end{cases}$$

The linear part of this vector field at $0 \in \mathbb{R}^2$ is nilpotent. One says that this singular point has *codimension 2*, because nilpotent, (2×2)-matrices form a codimension 2 subset in the space of all (2×2)-matrices. As a consequence, the germ of X_0 cannot be destroyed by a small perturbation when it appears in 2-parameter families of vector fields.

The following 2-parameter quadratic family unfolds $X_0^{N\pm}$:

$$X_\lambda^{N\pm} \begin{cases} \dot{x} = y \\ \dot{y} = x^2 + \mu + y(\mu \pm x) \end{cases} \qquad \text{where} \quad \lambda = (\mu, \nu) \in \mathbb{R}^2.$$

We will prove in the text that this unfolding of X_0 is *versal* in the sense that *any* other unfolding of X_0 factorizes through $X_\lambda^{N\pm}$ (up to some equivalence relation). As the two unfoldings $X_\lambda^{N\pm}$ are equivalent up to the orientation of trajectories and a linear change of parameter, we will only consider the unfolding X_λ^{N+}. It will simply be written X_λ^N.

Here we only want to describe the bifurcations of X_λ^N, and at the same time we give some heuristic idea of why all possible bifurcations of X_0 appear in the unfolding X_λ^N.

First of all, let us note that the phase portrait of X_0 near the origin is the same as the phase portrait of X_0^N, and that it represents an equilibrium point on a curve with a cusp shape passing through the origin. This is the reason why X_0 is often called the *Cusp singularity*. In fact, this phase portrait is the same as the phase portrait of the Hamiltonian vector field with Hamiltonian function $H_0 = \frac{1}{2}y^2 - \frac{1}{3}x^3$. Moreover, X_λ^N has the same singular points as the Hamiltonian unfolding $H_\mu = \frac{1}{2}y^2 - \mu x - \frac{1}{3}x^3$. For this Hamiltonian function, we obtain a pair of singular points (a center and a saddle point) where μ becomes negative. For the vector field belonging to the family, the center is replaced by a focus or a node (see Figure 2). The change of the center to a focus or a node can be explained by the effect of the second parameter ν which modifies the divergence of X_λ.

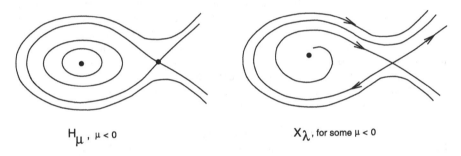

H_μ , $\mu < 0$ X_λ , for some $\mu < 0$

Figure 0.2

We now want to search for the possible limit cycles of X_λ and their bifurcations along the limit periodic sets of the family. Note that any limit cycle must surround the focus point C_λ and leave the saddle point s_λ outside the disk it bounds. As a consequence we have only three possible limit periodic sets:

- the focus point e_λ;

- a limit cycle γ_λ;

- a homoclinic saddle connection Γ_λ at the saddle s_λ.

A rather profound result about the family X_λ is that the cyclicity of any limit periodic set is always equal to one. We will see that this result is obtained by different methods depending on the nature of the limit periodic set: an algebraic computation proves that a Hopf bifurcation can occur at the focus point; the limit cycles are studied using the properties of abelian integrals, and the cyclicity of the saddle connection is obtained using asymptotic properties of a non-differential return map. As a consequence of this cyclicity property it is easy to see that for any $\mu < 0$, a unique unstable limit cycle is created by a *Hopf bifurcation* when ν crosses some value $\nu_H(\mu)$ in a decreasing direction. This limit cycle grows bigger as ν decreases from this value $\nu_H(\mu)$, and finally reaches the saddle point and creates a *saddle connection* for some value $\nu_C(\mu) < \nu_H(\mu)$. No limit cycles exist for ν outside the interval $[\nu_C(\mu), \nu_H(\mu)]$.

These considerations permit us to draw the bifurcation diagram of Figure 0.3. We observe that there are three lines of bifurcations: the axis $\{\mu = 0\}$, which is a line of saddle-node bifurcations SN along which a pair of singular points bifurcates, and two curves H and C (graphs of the functions $\mu \longrightarrow \nu_H(\mu), \nu_C(\mu)$ respectively, with $\mu < 0$), branching at the origin of the parameter space. These curves have a quadratic contact with SN. There exists one and only one limit cycle in the tongue between the two curves H, C. It is clear from the above description of the behaviour of the limit cycles in a function of the variation of the parameter, that the curve H is a line of Hopf bifurcations and C is a line of saddle connections. We can summarize this bifurcation diagram by saying that it explains how a small limit cycle is created near the equilibrium point and dies at the saddle connection, when the parameter varies. It is also interesting to note that this limit cycle is unstable and is the boundary of a small basin of attraction of a stable focus point. This small basin of stability disappears at the Hopf bifurcation, but is transformed into a thin region between separatrices of the saddle point, when the limit cycle arrives to the saddle connection (see Figure 0.3).

The study of bifurcations in any unfolding X_λ follows the same general lines. First one has to look for a possible model given by a polynomial family X_λ^N. This model family ($X_\lambda^{N\pm}$ in our example) is obtained, in general, using the *theory of normal forms* (see for instance [CLW], [D2], for general information about normal forms) and a suitable truncation of this normal form to a *sufficient finite jet*. Next, we study the bifurcation of singular points in our model family X_λ^N. As mentioned above, as X_λ^N is a polynomial family, this study can be made using algebraic algorithms. We return to the initial family X_λ by using implicit function theorem arguments. Finally, we study the limit cycles of the family, looking first for the possible limit periodic sets, and then computing their cyclicity and studying their local bifurcations. Sometimes, this reduces again to algebraic algorithms. This is the case for instance for the "Hopf-Takens" bifurcations of limit cycles at a weak focus. However, in general, the study of bifurcations of limit periodic sets is more involved. Several different methods are needed from analytic geometry, the theory of abelian integrals, the theory of asymptotics of some special classes of non-

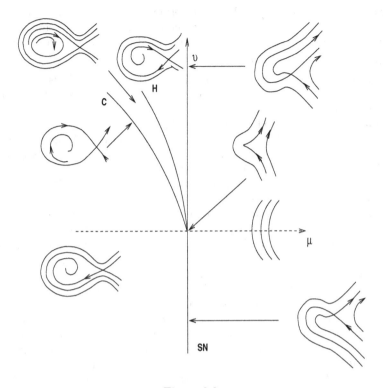

Figure 0.3

differentiable function families, and desingularization theory. *The aim of this text is to present these methods in a systematic way:* first, the analytic geometry used for *regular limit periodic sets*, next, the asymptotic methods which are related to the study of *elementary limit periodic sets*, and finally, the desingularization method applied to reduce *non-elementary limit periodic sets* to elementary ones.

We will not present a systematic review of possible generic bifurcation diagrams in functions of the number of parameters. Such a presentation may be found easily in the literature (see for instance [A], [GH], [CLW], [D], [DRS1], [DRS2], [KS]). The bifurcations which are presented in the text are given just to illustrate the methods. This is the case for the Bogdanov-Takens bifurcation which will be encountered several times, and for the bifurcations of the cuspidal loop, which is presented in the last chapter to illustrate the desingularization method.

An important application of the text is Hilbert's sixteenth problem. We have already mentioned that this problem can be replaced by a more general conjecture about the finite cyclicity of analytic unfoldings of limit periodic sets. This idea motivates the study of all possible limit periodic sets in polynomial families and then the study the cyclicity of these "polynomial" limit periodic sets. In Chapter 2, we give the results of this scheme for the special case of quadratic vector fields.

The plan of the text is as follows.

In Chapter 1, we recall some general properties of vector fields on surfaces of genus 0, such as the Poincaré-Bendixson theorem, and we make a first approach to their bifurcation theory.

Limit periodic sets, which are the main subject of the text, are introduced in Chapter 2, where their general properties are established. We prove a first and partial result about their structure, in the spirit of the Poincaré-Bendixson theory, and define the cyclicity $Cycl\ (X_\lambda, \Gamma)$. We show that Hilbert's problem reduces to a general conjecture on the finite cyclicity for analytic unfoldings. We describe this reduction in detail for quadratic vector fields.

In Chapter 3, we consider a single vector field (the "0-parameter case") and review the desingularization theory and the solution of the Dulac problem asking if an analytic vector field on S^2 has only a finite number of limit cycles. The reason for this survey is that some ideas that we introduce in Chapter 3 will be extended to families in the next chapters. For the Dulac problem, we restrict ourself to the case of hyperbolic graphics which can be easily studied and generalized to families.

The last three chapters are devoted to the study of bifurcations of limit periodic sets and to partial proofs of the conjecture on finite cyclicity. In Chapter 4, we consider regular limit periodic sets, i.e., elliptic singular points and periodic orbits. They can be studied by standard methods of differential and analytic geometry. Of particular interest is the notion of the Bautin ideal associated with analytic unfoldings of ∞-codimension vector fields of "centre type". This ideal, first introduced by Bautin for the study of polynomial quadratic vector fields, is extended here to general analytic unfoldings. Using it, one can obtain an estimate for the cyclicity. We also relate it to Melnikov functions and give some applications to quadratic vector fields.

In Chapter 5, we consider elementary graphics which are the limit periodic sets whose singular points are elementary. We prove a general structure result for the unfoldings of the transition map near an hyperbolic saddle. This result, together with the notion of the Bautin ideal, allows a complete study of the saddle connection unfoldings. In the last part of the chapter, we present a review of some recent results by Mourtada, El Morsalani, Il'Yashenko, Yakovenko and others about general elementary graphics.

The last chapter is devoted to non-elementary limit periodic sets and their desingularization. We treat the simplest case, a connection at a Bogdanov-Takens cusp point, in detail and give some indication about a general desingularization theory for families of vector fields.

The principal aim of this text is to bring together results from previous articles, although some results appear here for the first time. For instance, Theorems 4.2 and 5.13 give explicit bounds for the cyclicity in terms of the Bautin ideal and are applied in theorem 5.14 to quadratic vector fields. Section 6.2 presents results on generic bifurcations of the cuspidal loop. More detailed results about the cuspidal loop will appear in a forthcoming publication.

This text was prepared to support the lectures given in I.M.P.A. during the 20th Colóquio Brasilerio de Matemática", in July 1995. I wish to thank the Institution and the organizers of the Colloquium for their kind invitation. I want to thank my friend Jorge Sotomayor for his attentive reading of the manuscript and the numerous corrections he suggested. The text was typed in a short period in Dijon by Laurence Pidoux. I want to thank her for her efficiency and kindness.

The present version for publication has been corrected to take into account the referee's observations. I express my special acknowledgment to Pavao Mardešić for his many helpful suggestions.

Chapter 1
Families of Two-dimensional Vector Fields

1.1 Vector fields on surfaces of genus 0

In this section we will consider individual vector fields. They can be considered as 0-parameter families. We assume these vector fields to be of class at least $\mathcal{C}^1$. This will be sufficient to ensure the existence and uniqueness of the flow $\varphi(t,x)$ (t is time, $x \in S$, the phase space) and the qualitative properties which we mention below.

1.1.1 The phase space

We will suppose S to be a connected surface of genus 0, i.e., a submanifold of the 2-sphere S^2. This surface may or may not be compact. In the first case it may be S^2 or a surface with boundary, like the closed annulus $S^1 \times [0,1]$. In this case we will assume that the vector field is tangential or transversal to each component of the boundary ∂S. Later on, we will also consider boundaries with corners, such as the square $[0,1] \times [0,1]$, with some natural conditions for X along the boundary.

1.1.2 The Poincaré-Bendixson property

The principal reason for restricting ourselves to a phase space of genus 0 is to avoid nontrivial recurrence.

Nontrivial recurrence appears, for instance, if we consider the vector field $X = \dfrac{\partial}{\partial\theta_1} + \alpha\dfrac{\partial}{\partial\theta_2}$ on the 2-torus $\mathbb{T}^2 = \mathbb{R}^2/_{\mathbb{Z}^2}$, with angular coordinates (θ_1,θ_2); each orbit of X is dense on $\mathbb{T}^2$ when α is irrational. The ω-limit set of each orbit is the whole phase space. In contradistinction, for a 0-genus phase space, the following theorem holds:

Theorem 1 *(Poincaré-Bendixon) Let X be a vector field on a compact surface S of genus* 0.

1

R. Roussarie, *Bifurcations of Planar Vector Fields and Hilbert's Sixteenth Problem*, Modern Birkhäuser Classics, DOI: 10.1007/978-3-0348-0718-0_1, © Springer Basel 1998

Suppose that each singular point of X is isolated. Let γ be an orbit of X. Then its ω-limit set ω_γ is non-empty and falls into one of the following three cases:

1) *ω_γ is a singular point,*
2) *ω_γ is a periodic orbit,*
3) *ω_γ contains a non-empty subset of singular points Σ and at least one regular orbit. The ω and α-limit set of each regular orbit in ω_γ is one point of Σ.*

One can find a proof of the Poincaré-Bendixson theorem in [MP]. The basic idea of the proof consists of a very simple and natural topological argument, though some details are a bit delicate. As this argument will be useful later for studying families, we give it in the following lemma:

Lemma 1 *Let X be a vector field on a surface of genus 0. (We do not suppose that the singular points are isolated). Let γ be any orbit and $\sigma \subset S$ any interval transversal to X (i.e., for any $x \in \sigma$, the vector $X(x)$ is transversal to the tangent space $T_x\sigma$). Then $\omega_\gamma \cap \sigma$ contains at most one point.*

Proof. Suppose that $\omega_\gamma \cap \sigma$ contains at least two distinct points a, b. Let x be a point of γ. Consider the trajectory $\varphi(t, x)$. As $a, b \in \omega_\gamma$, it is possible to find three instants $t_1 < t_2 < t_3$ such that

1) $a_1 = \varphi(t_1, x) \in]a, b[$ (interval on σ with end points a, b),
2) Let σ_a and σ_b be the two open subintervals on σ with a_1 as one end point and a, respectively b, as the other. Then $a_2 = \varphi(t_2, x)$ is the first return of $\varphi(t, a_1)$ on σ_a,
3) $a_3 \in \sigma_b$.

Let Γ be the simple topological closed curve, defined as the union of the interval $[a_1, a_2]$ and the orbit arc $\varphi([t_1, t_2], x)$. Now, as S is a surface of genus 0, the curve Γ separates it into two connected components A and B, by Jordan's theorem.

One can choose A to be the component which contains the points $\varphi(t, a_2)$ for all $t > 0$, and B the component which contains points $\varphi(t, a_3)$ for all $t < 0$. This is a contradiction, because we can no longer go from a_2 to a_3 by the flow of X.   □

As a consequence of the lemma, any recurrent orbit cuts any transversal interval σ in at most one point. This is the case for a periodic orbit. Hence any recurrent orbit is a singular point or a periodic orbit: *on a surface of genus 0, a vector field has just trivial recurrence.*

If S is compact and if, as in Theorem 1, all singular points of X are isolated, then these singular points are finite in number. Moreover, if X is analytic, the number of regular orbits which occur in item 3 of the theorem must be finite. We will prove this point in Chapter 2, using the Desingularization Theorem. The third case in the theorem reduces to a graphic, according to the following definition:

Definition 1 *Let X be a vector field on a surface. A graphic Γ for X is a compact, non-empty, invariant subset which consist of a finite number of isolated singular points $\{p_1, \ldots, p_s\}$ and regular orbits $\{\gamma_1, \ldots, \gamma_\ell\}$ such that the ω and α-limit set of each of these regular orbits is one of the singular points. Moreover, Γ is the direct image of an S^1-immersion, oriented by increasing time.*

Finally, if Γ is the ω-limit set of a trajectory γ, one can choose a segment $\sigma \simeq [0,1[$ transversal to the vector field, with $0 \in \Gamma$, on one side of γ, such that a return map P is defined on σ, with $P(0) = 0$. We will say that Γ is a *monodromic graphic*.

We can now give a more precise formulation of the Poincaré-Bendixson theorem in the analytic case:

Theorem 2 *(Poincaré-Bendixson theorem for C^ω vector fields). Let X be an analytic vector field with isolated singular points on some compact surface of genus 0. Let γ be an orbit. Then ω_γ is a singular point, a periodic orbit or a monodromic graphic.*

1.1.3 Phase portrait

A consequence of the Poincaré-Bendixson theory is that, in general, a vector field on a surface of genus 0 admits a simple classification of its topological type. To make this precise, we need the notion of equivalence (of topological type) of vector fields.

Definition 2 *Let X, Y be two vector fields on S. They are* topologically equivalent *if there exists a homeomorphism h from S onto S, which sends each orbit of X onto an orbit of Y, preserving the time-orientation.*

If X and Y are topologically equivalent, one says that they have the same phase portrait. *(In fact one can identify the phase portrait of X with its equivalence class.)*

It seems hopeless to look for a classification of all possible phase portraits. To obtain interesting and useful results, one has to eliminate the too degenerate vector fields. So, one considers either generic smooth vector fields, or polynomial or analytic vector fields.

1.1.3.1 Generic vector fields. See [MP] for details.

To simplify, we suppose now that S is compact. We denote $\mathcal{X}^r(S)$, for $1 \leq r \leq \infty$, the space of all vector fields on S, with the usual C^r-topology of uniform convergence. As we are interested in this section in individual vector fields, it is of no interest to study vector fields whose phase portraits may be destroyed by an arbitrarily small perturbation. Hence the only vector fields of interest will be the stable ones, according to the following definition:

Definition 3 *Let $X \in \mathcal{X}^r(S)$. We say that X is C^s-structurally stable (for some $s : 1 \leq s \leq r$) if and only if there exists a neighborhood $\mathcal{U}$ of X in $\mathcal{X}^r(S)$, for the C^s-topology, such that any $Y \in \mathcal{U}$ is topologically equivalent to X.*

An important class of vector fields is defined as follow:

Definition 4 *$X \in \mathcal{X}^r(S)$ is said to be a Morse-Smale vector field if and only if*
  1) *all critical elements of X are hyperbolic (a critical element is a singular point or a periodic orbit),*
  2) *there is no trajectory connecting saddle points of X.*

The principal result for generic vector fields is:

Theorem 3 *(Andronov-Pontryagin, Peixoto). Let $1 \leq r \leq \infty$. Then*
  1) *the set $MS^r(S)$ of all Morse-Smale vector fields in $\mathcal{X}^r(S)$ is open and dense in $\mathcal{X}^r(S)$,*
  2) *a vector field $X \in \mathcal{X}^r(S)$ is C^r-structurally stable if and only if it is in $MS^r(S)$.*

Each Morse-Smale vector field has a finite number of critical elements. The singular points are hyperbolic sources, sinks or saddles. The periodic orbits may be hyperbolic attractive or repelling. It is possible, but rather tedious, to give a complete classification of the possible phase portraits.

I recall that the theorem of Andronov-Pontryagin and Peixoto was extended in 1962 by M. Peixoto to oriented compact surfaces of arbitrary genus M [P]. The proof in this case is much harder, due to the difficulty in eliminating nontrivial recurrence by small perturbations. This can be made by a C^r-perturbation for orientable M. But for non-orientable M, one needs a C^1-perturbation. The reason is that in this last case, it is not possible to construct a C^r-perturbation directly and one has to use the closing-lemma which has only been proved for the C^1-topology. See [MP] for a complete discussion and proofs of these difficult questions. On the other hand, the proof that any Morse-Smale vector field on M is C^r-structurally stable is very similar to the proof for surfaces of genus 0.

1.1.3.2 Analytic and polynomial vector fields

Polynomial vector fields are of particular interest, even if at first sight they seem to form a rather special class of vector fields. In fact, their popularity in many fields of application (electrotechnics, ecology, biology, . . .) comes from their apparent simplicity, and as a consequence the belief that their properties would also be simple. We will see that this belief is somewhat misleading. Nevertheless, one clear advantage of polynomial vector fields is that they are easy to integrate numerically and hence, they lead to simple models.

Another important point is that polynomial vector fields of a given degree n form an explicit family with a finite number of parameters. We call it $\mathcal{P}_n$. Some questions which make no sense in the general context of generic vector fields can now be addressed: Is it possible to find algorithms to locate singular points and periodic orbits? Is it possible to search for them by means of an algorithm which can be implemented on a computer? We will come back to these questions in the context of families of vector fields.

Of course one can study the problems of Morse-Smale vector fields inside the space $\mathcal{P}_n$, endowed with the topology of coefficients. The density question was solved by J. Sotomayor in [So2]. But it appears that the question whether *each structurally stable polynomial vector field is a Morse-Smale one* remains unsolved in general. Moreover, it will be essential for the study of generic smooth unfoldings to consider all polynomial vector fields and not just the Morse-Smale ones, as would be natural in the generic context for individual vector fields. For instance, the unfoldings of Hamiltonian vector fields will be of interest.

Almost all the questions about polynomial vector fields that we will address only make use of their analyticity.

We will just use the fact that any polynomial vector field X on $\mathbb{R}^2$ extends to an analytic vector field $\widetilde{X}$ on S^2. The most straightforward way to obtain it is given by the following *Bendixson compactification*.

Identifying $\mathbb{R}^2$ with $\mathbb{C}$ by taking $x + iy = z$, one can write the differential equation of $X \in \mathcal{P}_n$ as

$$\dot{z} = P(z, \bar{z}) = \sum_{0 \le i+j \le n} a_{ij} \, z^i \bar{z}^j \, , \text{ with } a_{ij} \in \mathbb{C} \tag{1.1}$$

Let $Z = 1/z$ be the chart at infinity in $\mathbb{C}$ (the two charts $z \in \mathbb{C}$ and $Z \in \mathbb{C}$ form an atlas of $S^2 = \mathbb{C} \cup \{\infty\}$).

In this chart, the equation (1.1) becomes:

$$\dot{Z} = -\frac{\dot{z}}{z^2} = -Z^2 \sum a_{ij} \, Z^{-i} \, \overline{Z}^{-j} \tag{1.2}$$

Of course, the vector field tends to infinity when $Z \to 0$ or $z \to \infty$. But if we multiply X by the real analytic function $f_n(z) = \dfrac{1}{1 + (z\bar{z})^n}$, it is easy to see that the vector field $\widetilde{X} = f_n X$ is analytic on S^2. This is clear in the chart z. In the chart Z, $\widetilde{X}$ is given by:

$$\dot{Z} = -\frac{Z^2}{1 + (Z\overline{Z})^{-n}} \sum_{0 \le i+j \le n} a_{ij} Z^{-i} \overline{Z}^{-j} = -\frac{Z^2 (Z\overline{Z})^n}{1 + (Z\overline{Z})^n} \sum a_{ij} Z^{-i} \overline{Z}^{-j}$$

$$\tag{1.3}$$

$$\dot{\mathbb{Z}} = -\frac{Z^2}{1 + (Z\overline{Z})^n} \sum_{0 \le i+j \le n} a_{ij} \, Z^{n-i} \overline{Z}^{n-j}$$

which is also analytic in Z.

In this compactification, just one point is added to $\mathbb{R}^2$. In the *Poincaré compactification*, a circle of points is added to $\mathbb{R}^2$ (the circle at infinity). One obtains a vector field $\widetilde{X}_1$ on D^2, and $\mathbb{R}^2$ is identified to $\mathrm{int}(D^2)$. To obtain this vector field $\widetilde{X}_1$ from $\widetilde{X}$, it will suffice to blow-up the point ∞ of S^2. This blow-up procedure will be explained in Chapter 3.

In Chapter 3, we will also return in detail to the study of individual polynomial and analytic vector fields.

1.2   A first approach to Bifurcation Theory

In general, a vector field will depend on parameters.

For a vector field modelling some natural phenomena, the parameters will represent the exterior world: damping coefficients, frequencies, – in mechanics; growth rates, rates of predation, – in ecology; etc. Hence, even when one just wants to consider an individual vector field, the parameters will in fact be present in an implicit way.

For instance, if we consider a given polynomial vector field X_0 of degree n, it will be natural to look at its coefficients as parameters. What happens when we modify the coefficients of X_0 is of utmost importance in deciding on the interest of modelling of a phenomenon by the explicit vector field X_0.

The embedding of the vector field X_0 inside a whole family X_λ raises many interesting new bifurcation problems. Their study is precisely the subject of these notes. In this section, we will introduce some general definitions, which can easily be extended to a more general context, for dynamical systems in any dimension. In the next chapter, we will introduce some notions more specific to families of vector fields on surfaces.

A family of vector fields (X_λ) on S is a map of a parameter space P to the space $\mathcal{X}^r(S) : \lambda \to X_\lambda$.

Here, $1 \leq r \leq \infty$ or $r = \omega$ for analytic families. We will also consider the previously mentioned family $\mathcal{P}_n$ of vector fields of degree $\leq n$. For this family the parameter is the set of coefficients. It is the Euclidean space $\mathbb{R}^{N(n)}$, with $N(n) = (n+1)(n+2)$. In general, the parameter space P will be a manifold of finite dimension p, usually $\mathbb{R}^p$. We will say that the family (X_λ) is $\mathcal{C}^k$ if it is given in a neighborhood of each $(x_0, \lambda_0) \in S \times P$ by the expression

$$X_\lambda(x, \lambda) = \sum_{i=1}^{n} a_i(x, \lambda) \frac{\partial}{\partial x_i} \, ,$$

where $a_i(x, \lambda)$ are $\mathcal{C}^k$ functions defined on $U_{x_0} \times W_{\lambda_0}$. Here U_{x_0} and W_{λ_0} are the charts of S and P respectively in a neighborhood of x_0 and λ_0 and $k = 1, \ldots, \infty$ or ω.

Of course, one can see a $\mathcal{C}^k$-family (X_λ) as a vector field defined on the total space $S \times P$, which is tangent to the fibers of the projection $\pi : S \times P \to P$,

$(\pi(x, \lambda) = \lambda)$; i.e., (X_λ) is associated to the field $X(x, \lambda) = X_\lambda(x)$ on $S \times P$ which verifies $d\pi(x, \lambda)[X(x, \lambda)] \equiv 0$.

If we denote by $\chi^r(S; P)$ the space of all C^r-families of vector fields with parameter belonging to P, then the previous remark means $\chi^r(S; P) \subset \chi^r(S \times P)$. We endow $\chi^r(S; P)$ with the topology induced by the C^r-topology on $\chi^r(S \times P)$.

Let (X_λ) be a family. The set in the parameter space where X_λ is structurally stable is an open set $U(X_\lambda)$. On each connected component of $U(X_\lambda)$, the phase portrait of X_λ is constant. By contrast, this phase portrait changes in general when we cross $\Sigma(X_\lambda) = P - U(X_\lambda)$. This is the reason why one calls this set *the bifurcation set* of the family (X_λ). This set is closed. It seems reasonable to think that this set has an empty interior, in general. In fact, this is false as soon as $\dim S \geq 3$, as proved by Smale [Sm]. (Smale considered the whole family $\chi^r(S)$, where the parameter space is the space $\chi^r(S)$ itself, and exhibited an open set of non-structurally-stable vector fields; it follows that there exist generic finite parameter families with any number of parameters, such that the bifurcation set has a non empty interior.) When $\dim S = 2$, the question whether the parameter space is larger than two remains open.

One knows, however, that structural stability is a generic property, and that $\Sigma(X_\lambda)$ has generically an empty interior for 1-parameter families. This was proved by Sotomayor in [So1]. If Σ is the set of all non-structurally-stable vector fields in $\chi^r(X)$, then $\Sigma((X_\lambda)) = \rho^{-1}(\Sigma)$ where ρ is the map $\lambda \to (X_\lambda)$. The structure of the sets $\Sigma((X_\lambda))$ depends on the structure of Σ. If one could apply the transversality theorem to Σ (i.e., if Σ would be a stratified set, with finite codimension strata), then for each generic family X_λ, $\Sigma((X_\lambda))$ would also have the structure of a stratified subset of S. In particular, $\Sigma((X_\lambda))$ would be generically nowhere dense. But to establish this property for Σ looks rather utopian. Up to now, only the (codimension 1)-skeleton of Σ [So1] is known. Describing the 2-skeleton seems rather complicated. Of course the study of the skeleton of Σ, up to codimension k, is completely equivalent to the study of all generic families with less than k parameters. In these notes we want to prove some very partial results in these direction. In fact, we will emphasize some particular aspects of the bifurcation theory especially concerning the number of isolated periodic orbits, i.e., the question of *finite cyclicity*. We introduce this concept of cyclicity in the next chapter. Next some basic and general definitions about families of vector fields are given.

1.2.1 General definitions

Definition 5 *Let $(X_\lambda), (Y_\lambda)$ be two C^r families of vector fields with the same parameter space P and the same phase space S; $r = 1, \ldots, \infty$, or ω and let $s : 0 \leq s \leq r$. We say that (X_λ) and (Y_λ) are $(C^0$-fibre, $C^s)$-equivalent, if there exists a diffeomorphism φ of P, of class C^s, such that for each $\lambda \in P$, X_λ and $Y_{\varphi(\lambda)}$ are topologically equivalent. If the equivalence homeomorphism may be chosen so that it forms a continuous family $h_\lambda(x)$, one says that (X_λ) and (Y_λ) are (C^0, C^s)-*

equivalent. *If $\varphi = Id$, one says simply that (X_λ) and (Y_λ) are respectively C^0-fibre or C^0-equivalent.*

Remark 1

1) *The (C^0, C^s)-equivalence may be strengthened to the notion of (C^ℓ, C^s)-equivalence or even conjugacy (for $0 \leq \ell, s \leq r$). But, these equivalences are in general too strict to avoid problems of moduli between generic families which are $(C^0$-fibre, $C^s)$-equivalent. For instance, this occurs for the Hopf-Takens bifurcations with more than 4 parameters (see [R1]).*

2) *It will be important to have some smoothness for the change of parameter φ, since we are interested in the differentiable or even analytic structure of the bifurcation set: for applications it will be relevant to know if 2 lines of bifurcation have some flat contact for instance.*

Definition 6 *Let (Y_λ), $\lambda \in P$ be a C^r-family and $\varphi : Q \to P$, $\varphi(\mu) = \lambda$, be a C^r-map. We say that the family $X_\mu, \mu \in Q$, given by $X_\mu = Y_{\varphi(\mu)}$ is induced from Y_λ by the map φ.*

Remark 2 *This operation of induction will be very important in the selection of a "good" set of parameters, i.e., for replacing a parameter μ by a larger one λ, where the properties are "unfolded". (See the notion of versal unfolding below.)*

Definition 7 *A germ of a family (X_λ, λ_0) at $\lambda_0 \in P$ is called an* unfolding *of X_λ at λ_0. More generally, if $\Gamma \subset S$ is a compact non-empty invariant subset of X_{λ_0}, we will also consider the germ of X_λ along $\Gamma \times \{\lambda_0\}$. We will abbreviate this unfolding by (X_λ, Γ), or by $(X_\lambda, \Gamma, \lambda_0)$ if we want to recall the parameter value.*

Unfoldings are represented by local families. For instance, (X_λ, Γ) will be represented by a family $\widetilde{X}_\lambda$ on some neighborhood $U \times W$ of $\Gamma \times \{\lambda_0\}$ in $S \times P$. We can always assume that W is diffeomorphic to $\mathbb{R}^p$ ($p = \dim P$), and U is diffeomorphic to $\mathbb{R}^2$, if Γ is a point. So it will be synonymous to speak of unfoldings or local families, i.e., families defined on an arbitrary neighborhood of $\Gamma \times \{\lambda_0\}$.

Remark 3 *The two preceding definitions of equivalence and induction are easily translated for unfoldings, by taking representative local families.*

Definition 8 *Let Γ be a compact non-empty invariant subset for X_{λ_0}. We say that (X_λ, Γ) is a* versal unfolding *for the germ (X_{λ_0}, Γ) for the $(C^0$-fibre, $C^s)$-equivalence, if*

1) *Any other unfolding (Y_μ, Γ) of (X_{λ_0}, Γ) (i.e., any unfolding (Y_μ, Γ) with $\mu \in Q$ a parameter space, and Γ an invariant set for Y_{μ_0} where $(Y_{\mu_0}, \Gamma) \equiv (X_{\lambda_0}, \Gamma)$) is C^0-fibre equivalent to an unfolding induced from X_λ by a germ of a C^s-map $(\varphi, \mu_0) : (Q, \mu_0) \to (P, \lambda_0)$.*

2) *$\dim(P)$ is minimal for the property 1.*

1.2.2 Singularities of finite codimension. The saddle-node bifurcations

We refer the reader to [Ma], [GG] for details about transversality theory and notions related to it: genericity, codimension of a singularity, versality and structural stability of unfoldings. Here we just give a brief survey of the terminology.

We define a *singularity* Σ *of codimension* k as a submanifold of codimension $n + k$ (if $n = \dim S$) of some space of l−jets of vector fields on S. This submanifold is supposed to be invariant under the natural action of $(l + 1)$-jets of diffeomorphisms of S. A germ (X, x_0) is "a singularity of type Σ" if $j^l X(x_0) \in \Sigma$. Now, a consequence of the Thom's Transversality Theorem is that generically any C^{l+1} family of vector fields has an l-jet extension $(x, \lambda) \in S \times P \to j^l X_\lambda(x)$, transversal to Σ. This implies that the set $\Sigma(X_\lambda) = j^l X_\lambda^{-1}(\Sigma)$ is a submanifold of codimension $k + n$ in $S \times P$, and that in particular $\Sigma(X_\lambda) = \emptyset$, if $k > p = \dim P$.

Any known versal family is constructed in the following way. The germ (X, x_0) is taken as some singularity Σ of codimension k and the versal unfolding $(X_\lambda, \{x_0\})$ of (X, x_0) is any k-parameter unfolding whose jet-extension is transversal to Σ. In all known cases, the versal unfolding $(X_\lambda, \{x_0\})$ is also *structurally stable* in the following sense: any nearby k-parameter unfolding (Y_λ) has a jet-extension cutting Σ at some (x_1, λ_1) close to (x_0, λ_0), and the germ of $(Y_\lambda, \{x_1, \lambda_1\})$ is equivalent to the germ $(X_\lambda, \{x_0, \lambda_0\})$. This implies also that all the germs on Σ, near (X, x_0), are equivalent to (X, x_0). If Σ is defined in the space of K-jets, we can take in particular any germ with the same K-jet as (X, x_0): this means that any germ of Σ near (X, x_0) is K-*determinate*.

In general one reverses the terminology: singularity of finite codimension k means singularity whose transversal unfoldings are versal and structurally stable. Then the singularity is finitely determinate and the codimension k is equal to the number of parameters of any versal unfolding.

A pragmatic way to construct a versal family is as follows: one selects a (finite determinate) singularity of codimension k and considers any transversal unfolding to it. This procedure is inspired by the unfolding theory of differentiable maps, as developed by Thom, Arnold, Mather and others. Unfortunately, there are no general results in this direction for unfoldings of vector fields (even on surfaces).

To illustrate the above remarks, let us consider *the saddle-node bifurcations of codimension* k, $k \geq 1$.

We say that a germ $(X, 0)$ at $0 \in \mathbb{R}^2$ is a saddle-node singularity of codimension k, $k \geq 1$ if

1) $j^1 X(0)$ has only one eigenvalue equal to 0,

2) Let W be any center manifold through 0, of class larger than k, then

$$X \mid_W (x) = [\alpha x^{k+1} + O(x^{k+1})] \frac{\partial}{\partial x}$$

with $\alpha \neq 0$ where x is a parametrization of W, with $x = 0$ at the origin.

It is easy to verify that the set of all germs $(X, 0)$ with these properties defines a submanifold $SN(k)$ in the $(k + 1)$-jet space of vector fields at $0 \in \mathbb{R}^2$.

This submanifold is of codimension $k+2$. Using the Center Manifold Theory [CLW] and the Preparation Theorem [M], it is possible to prove that any $\mathcal{C}^\infty$ unfolding $(X_\lambda, 0)$ of $(X, 0)$, with $\lambda_0 = 0$, is, for any finite $l >> k$, $\mathcal{C}^l$-equivalent (i.e., by a $\mathcal{C}^l$ family of diffeomorphisms and multiplicative functions) to

$$\pm y \, \frac{\partial}{\partial y} \, \pm \left[x^{k+1} + \sum_{i=0}^{n-1} \alpha_i(\lambda) x^i \right] \frac{\partial}{\partial x}$$

where $\alpha_i(\lambda)$ are $\mathcal{C}^l$ germs of functions at $0 \in \mathbb{R}^p$.

This means that the unfolding

$$Y_\alpha = \pm y \, \frac{\partial}{\partial y} \pm \left[x^{k+1} + \sum_{i=0}^{n-1} \alpha_i \, x^i \right] \frac{\partial}{\partial x} \ , \quad \alpha = (\alpha_0, \dots, \alpha_{k-1})$$

is a versal unfolding for $(X_0, 0)$ for the $(\mathcal{C}^l, \mathcal{C}^l)$-equivalence. Moreover, this unfolding is structurally stable and it is easy to verify that
 - any germ in $SN(k)$ is $(k+1)$-determinate,
 - the $(k+1)$-jet extension of Y_α is transversal to $SN(k)$.

One can find the details of the proofs and the corresponding bifurcation diagrams in [D2]. Recent results for $\mathcal{C}^l$ conjugacy are given in [IY1].

1.2.3 Bifurcations of singular points versus bifurcations of periodic orbits. The Bogdanov-Takens bifurcation

What makes the study of saddle node bifurcations easy is that no periodic orbit is contained in a neighborhood of the origin. If we consider for instance a polynomial family of vector fields (X_λ), $\lambda \in \mathbb{R}^p$, then all the properties concerning singular points are described by polynomial equations or inequalities. Let $X_\lambda = A_\lambda \, \frac{\partial}{\partial x} + B_\lambda \, \frac{\partial}{\partial y}$, where A_λ, B_λ are polynomials of degree $\leq n$, in x, y, depending linearly on the parameter λ. The singular set is given by the polynomial equations

$$A_\lambda = B_\lambda = 0. \tag{1.4}$$

If we want to look at the set of degenerate singular points, we have to add the equation

$$\frac{\partial(A, B)}{\partial(x, y)} = 0 \tag{1.5}$$

The set of parameter values where one has at least one degenerate singular point is obtained by the elimination of (x, y) between (1.4), (1.5): it is a semi-algebraic subset in the parameter space. The set of parameters where X_λ has a saddle node point of codimension k is also a semi-algebraic set, defined as the projection on $\mathbb{R}^p$ of the semi-algebraic set $j^{k+1} X_\lambda^{-1} (SN(k))$, and so on.

On the other hand, if we want to study the periodic orbits, we have to integrate the vector field X_λ and the properties defined via the flow of X_λ are in general no longer given by algebraic algorithms, even if the family is polynomial. For instance, to study the periodic orbits cutting some line interval $\sigma \subset \mathbb{R}^2$, one proceeds as follow. Suppose that for some value $\lambda_0 \in P$ there exists some line interval σ', $\sigma' \supset \bar{\sigma}$, transversal to X_{λ_0} and such that for each $u \in \sigma$, the trajectory through u comes back the first time at the point $P_{\lambda_0}(u)$ on σ'. It follows by continuity that there exists a neighborhood W of λ_0 in P, and a first return map $P_\lambda(u) : \sigma \times W \to \sigma'$. Now, the key remark is that we just know that $P_\lambda(u)$ is analytic, and in general we cannot deduce more information from the fact that X_λ is polynomial. Hence, the equation $\{P_\lambda(u) - u = 0\}$ which gives the equation of pairs (u, λ) such that X_λ has a periodic orbit through u (i.e., the equation of periodic orbits cutting σ) is just analytic and, in general, there is no algebraic algorithm to solve it.

To illustrate this point, we present the Bogdanov-Takens bifurcation. This bifurcation is the most complex one of codimension 2, and we need to study it to have a complete list of all generic unfoldings with less than two parameters. We use it to illustrate the methods for the study of unfoldings along these lines. A complete treatment of this bifurcation can be found in [Bo], [T1], [RW].

Bogdanov-Takens bifurcation

Let $(X, 0)$ be a germ such that $j^1 X(0)$ is nilpotent, i.e., linearly conjugate to $y \dfrac{\partial}{\partial x}$. We can assume that $j^1 X(0) = y \dfrac{\partial}{\partial x}$. It is easy to show that, up to a quadratic diffeomorphism, one has

$$j^2 X(0) = y \frac{\partial}{\partial x} + ax^2 \frac{\partial}{\partial y} + b\,xy\,\frac{\partial}{\partial y}, \quad \text{with } a, b \in \mathbb{R}.$$

Generically we can suppose that $a \neq 0$ and $b \neq 0$. Then, by a linear change of coordinates we obtain $a = 1$ and $b = \pm 1$. Hence, one can suppose that

$$j^2 X(0) = y \frac{\partial}{\partial x} + x^2 \frac{\partial}{\partial y} \pm xy \frac{\partial}{\partial y}. \tag{1.6}$$

Germs $(X, 0)$ with 2-jets equivalent to (1.6) form a singularity of codimension 2 in the space of 2-jets of vector fields with two connected components TB^+, TB^- (depending on the sign $\pm$ of $xy \dfrac{\partial}{\partial y}$). We will see in Chapter 3 that the germ $(X, 0)$ has the same phase portrait as the Hamiltonian vector field $y \dfrac{\partial}{\partial x} + x^2 \dfrac{\partial}{\partial y}$, with a cusp singular point at the level through the origin. For this reason, $(X, 0)$ is called a *cusp singularity*.

Such a singular point is the simplest non-elementary singularity (see Chapter 3).

Now let (X_λ) be any C^∞ unfolding of $(X, 0)$, with $(X_0, 0) = (X, 0)$. (The unfolding is defined at $\lambda_0 = 0 \in \mathbb{R}^p$) with $(X_0, 0) \in TB \pm$. Let

$$\dot{x} = H_1(x, y, \lambda) = y + 0(\|m\|^2) + 0(|\lambda|) \ , \ \dot{y} = H_2(x, y, \lambda) \tag{1.7}$$

be the differential equation of (X_λ) ; $m = (x, y)$, $\|.\|$ a norm on $\mathbb{R}^2$. One has $\dfrac{\partial H_1}{\partial y}(x, y, \lambda) \neq 0$. Thus, we can take as local coordinates $Y = H_1(x, y, \lambda)$, $X = x$ as local coordinates. Renaming the coordinates by x, y, the equation (1.7) takes the form

$$\begin{cases} \dot{x} &= y \\ \\ \dot{y} &= F(x, y, \lambda) = g(x, \lambda) + yf(x, \lambda) + y^2 Q(x, y, \lambda), \end{cases}$$

where g, f and Q are C^∞ functions.

By hypothesis, $g(x, 0) = x^2 + 0(x^3)$, $f(x, 0) = \pm x + 0(x^2)$.

The systems is now equivalent to a second-order differential equation

$$\ddot{x} = g(x, \lambda) + \dot{x}f(x, \lambda) + \dot{x}^2 Q(x, \dot{x}, \lambda). \tag{1.8}$$

It contains a Hamiltonian part, $\ddot{x} = g(x, \lambda)$, corresponding to the function

$$H(x, y, \lambda) = \frac{1}{2} y^2 + G(x, \lambda)$$

where $G(x, \lambda) = -\int_0^x g(s, \lambda)ds$.

Now, as $G(x, 0) = \dfrac{1}{3} x^3 + 0(x^3)$, the Hamiltonian function $H(x, 0, 0)$ has a versal unfolding in the sense of Catastrophe Theory. Hence, there exists a C^∞ differentiable change of coordinates with parameter

$$x = U(X, \lambda) = X + 0(X) + 0(\lambda) \tag{1.9}$$

so that

$$-g(x, \lambda)dx = -(X^2 + \mu(\lambda))dX \tag{1.10}$$

for some C^∞ function $\mu(\lambda)$ (see [Ma]).

Using the C^∞ change of coordinates $x = U(x, \lambda)$, $y = y$, we obtain that (1.8) is differentiably equivalent (i.e., up a C^∞ diffeomorphism, and a multiplication by a C^∞ positive function) to

$$\begin{cases} \dot{x} &= y \\ \dot{y} &= x^2 + \mu(\lambda) + y\Big(\nu(\lambda) \pm x + x^2 h(x, \lambda)\Big) + y^2 Q(x, y, \lambda) \end{cases} \tag{1.11}$$

where $\mu(\lambda)$ and $\nu(\lambda)$ are C^∞ functions such that $\mu(0) = \nu(0) = 0$ and $h(x, \lambda)$, $Q(x, y, \lambda)$ are C^∞ functions.

At this point it seems interesting to choose μ, ν as new parameters. There are two ways to achieve this:

1) supposing that X_λ is a generic 2-parameter unfolding in the sense that $\frac{\partial(\mu,\nu)}{\partial(\lambda_1,\lambda_2)}(0) \neq 0$, then, by a $\mathcal{C}^\infty$ change of parameters, we can suppose that $\lambda = (\lambda_1, \lambda_2) = (\mu, \nu)$. Or,

2) introducing the new family

$$X^\pm_{\mu,\nu,\lambda} \begin{cases} \dot{x} &= y \\ \dot{y} &= x^2 + \mu + y\Big(\nu \pm x + x^2 h(x,\lambda)\Big) + y^2 Q(x,y,\lambda). \end{cases} \tag{1.12}$$

The initial family (1.8) is then induced by the $\mathcal{C}^\infty$ map $\mathcal{Q} : \lambda \to \big(\mu(\lambda), \nu(\lambda), \lambda\big) \in \mathbb{R}^{2+p}$.

If we prove that the parameter λ in (1.13) plays no role in the sense that there exists a $\mathcal{C}^\infty$ map of maximal rank $\pi(\mu, \nu, \lambda)$ on the space (μ, ν), with $\pi(0,0,0) = (0,0)$ such that for each (μ, ν, λ) near $(0,0,0)$, $X_{\mu,\nu,\lambda}$ is equivalent to $X^{N\pm}_{\pi(\mu,\nu,\lambda)}$ where

$$X^{N\pm}_{\mu,\nu} \begin{cases} \dot{x} &= y \\ \dot{y} &= x^2 + \mu + y(\nu \pm x). \end{cases} \tag{1.13}$$

then, the initial family (1.8) will be ($\mathcal{C}^0$-fibre, $\mathcal{C}^\infty$)-equivalent to the family induced by the map $\pi \circ \Phi$, and we will have proved the following:

Theorem 4 *(Bogdanov-Takens). The polynomial unfolding $X^{N\pm}_{\mu,\nu}$ is a versal unfolding of the cusp singularity (Bogdanov-Takens singularity defined by (1.6)) for the ($\mathcal{C}^0$-fibre-$\mathcal{C}^\infty$) equivalence.*

Remark 4 *It is possible to obtain a $(\mathcal{C}^0, \mathcal{C}^\infty)$-equivalence ([DR1]).*

Of course, if we just consider generic 2-parameter unfoldings, we obtain a weaker result: every generic two parameter unfolding of the cusp singularity is ($\mathcal{C}^0$-fibre, $\mathcal{C}^s$) equivalent to $X^{N\pm}_{\mu,\nu}$. But the result as given in the theorem is better in the sense that it applies to unfoldings with any dimension of parameter and, in particular, it is absolutely essential when studying unfoldings of singularities of $cod \geq 3$.

In order to begin the proof of Theorem 4, we return to the family unfolding (1.13). Note that if we change $(x, y, \mu, \nu, \lambda)$ to $(x, -y, \mu, \nu, \lambda)$, then the unfolding $X^+_{\mu,\nu,\lambda}$ changes into the unfolding $-X^-_{\mu,\nu,\lambda}$. Hence it suffices to look to the $+$ case. The equation for singular points is given by $y = 0$, $x^2 + \mu = 0$. There exist no singular points for $\mu > 0$, and two singular points for $\mu < 0$: $e_\mu = (-\sqrt{-\mu}, 0)$ and $s_\mu = (\sqrt{-\mu}, 0)$. It is easy to verify that s_μ is a saddle point, and e_μ a node or a focus. Moreover, the line $\{\mu = 0\}$ for $\nu \neq 0$ is a line of (codimension 1) saddle-node bifurcations: when we cross the axis 0ν, for $\nu \neq 0$, and in the direction of negative μ, a pair of singular points $\{e_\mu, s_\mu\}$ appears.

At any point $(x,0)$, div $X_\Lambda(x,0) = \nu + x + x^2 h(x,\lambda)$ (we put $\Lambda = (\mu,\nu,\lambda)$).
In particular, div $X_\Lambda(e_\mu) = \nu - \sqrt{-\mu} + x^2 h(x,\lambda)$.

Hence, the equation div $X_\Lambda(e_\mu) = 0$ defines (for any λ) a curve H : $\mu = \mu_h(\nu) = -\nu^2 + 0(\nu^2)$ for $\nu \geq 0$ in the + case and $\nu \leq 0$ in the − case. We only consider the + case from now on. We write $\nu = \nu_h(\mu)$, $\mu \leq 0$ for the inverse of $\mu_h(\nu)$. Along the line H, the singular point e_μ is elliptic (its eigenvalues are purely imaginary). In fact it is easy to prove that div $X_\Lambda(e_\mu)$ changes sign when crossing H and that H is a generic line of Hopf bifurcations of codimension 1 (see Chapter 4).

Crossing the line H with decreasing ν, the focus e_μ becomes stable, and a small unstable periodic orbit γ_Λ around e_μ appears. We do not study it for a moment to look at the left hand separatrices of the saddle point s_μ. If a small negative value $\mu_0 < 0$ is chosen, for a fixed value λ_0, it is easy to see that these two separatrices cross the $0x$ axis at points $a(\lambda)$ for the lower separatrix and $b(\nu)$ for the upper one. Now, observe that if ν decreases, then the vector $X_\Lambda(x,y)$, for $\Lambda = (\mu_0,\nu,\lambda_0)$, rotates in the positive direction, for each $(x,y) \in \mathbb{R}^2$ with $y \neq 0$. As a consequence, $\dfrac{da}{d\nu}(\nu) > 0$ and $\dfrac{db}{d\nu}(\nu) < 0$.

For small values of ν, one has $a(\nu) < b(\nu)$, and for large values, $a(\nu) > b(\nu)$. Then there exists only one value $\nu_0 = \nu_c(\mu_0)$ where

$$a(\nu_0) = b(\nu_0).$$

For this value ν_0 one has a saddle connection. Moreover,

$$\frac{d}{d\nu}\,(a - b)(\nu) \neq 0$$

and also

$$div X_{\Lambda_0}(s_{\mu_0}) \neq 0 \Big(\Lambda_0(\mu_0,\nu_0,\lambda_0)\Big).$$

This means that the saddle connection is a generic codimension 1 connection (see Chapter 5). When ν decreases from values greater than ν_0, a large periodic orbit exists, which for $\nu = \nu_0$ becomes the saddle connection and disappears for $\nu > \nu_0$. The curve $C = \{\nu_0 = \nu_c(\mu_0) \mid \mu_0 < 0$ small enough $\}$ is a generic line of codimension 1 saddle connections.

The two lines $H = \{\nu = \nu_h(\mu)\}$ and $C = \{\nu = \nu_c(\mu)\}$ form the boundary of a conic region T in the parameter space. A small periodic orbit appears in this region near H and a large one disappears near C. It seems reasonable to think that it is the same orbit which appears on one side and disappears on the other, and also that for each $(\mu,\nu) \in T$ we have just one periodic orbit.

Moreover, there is no periodic orbit outside T, in a whole neighborhood of the origin in the parameter space. This will give the complete description of the bifurcation diagram and is the essential part of the proof of Theorem 4 (see Figure 3 in Preface). But the proof of this point is unexpectedly delicate. The use of a rescaling in phase space and parameter space (see Chapter 6), asymptotic methods and a fine result on Abelian integrals (see Chapters 4, 5) are needed. The reason is that we have no simple algorithm to control the return map on $[e_\mu, s_\mu]$.

The above example illustrates the difficulties encounted in studying periodic orbits, even in the very simple family $X_{\mu,\nu}^{N\pm}$. We will return to it several times to illustrate different technical ideas and to achieve a complete proof of Theorem 4. In the next chapter, we focus on the principal subject of these notes, *how to study bifurcations of periodic orbits?*

Chapter 2
Limit Periodic Sets

As explained at the end of the previous chapter, the most difficult problem in the study of bifurcations in a family of vector fields on a surface of genus 0 is the control of the periodic orbits. In fact, in generic smooth families the periodic orbits will be isolated for each value of the parameter. For analytic families we have two possibilities for each orbit: it may be isolated or belong to a whole annulus of periodic orbits. In this last case and for the parameter values for which the system has infinitely many periodic orbits, the vector field has a local analytic first integral and the nearby vector fields in the family may be studied by the perturbation theory introduced in Chapter 4. They have in general isolated periodic orbits. The interest in the study of isolated periodic orbits is also justified by tradition and by applications.

Definition 9 *A* limit cycle *of a vector field X in dimension 2 is a periodic orbit γ which is isolated on one side, i.e., not approached by periodic orbits, all belonging to one side of γ. (If X is analytic a limit cycle is necessarily isolated on both sides.)*

The most famous question about limit cycles was formulated by D. Hilbert in his inaugural talk at the first International Congress of Mathematicians in Paris (1901). The 16th problem in the list he submitted to the audience had a "part a" about the classification of ovals defined by a polynomial equation $\{H(x, y) = 0\}$, and a "part b" about the limit cycles of polynomial vector fields. Let us quote this part b of Hilbert's 16th problem:

> ... *In connection with this purely algebraic problem, I wish to bring forward a question which, it seems to me, may be attacked by the same method of continuous variation of coefficients, and whose answer is of corresponding value for the topology of families of curves defined by differential equations. This is the question as to the maximum number and position of Poincaré's boundary cycles (limit cycles)*

R. Roussarie, *Bifurcations of Planar Vector Fields and Hilbert's Sixteenth Problem*, Modern Birkhäuser Classics, DOI: 10.1007/978-3-0348-0718-0_2, © Springer Basel 1998

for a differential equation of the first order of the form

$$\frac{dy}{dx} = \frac{Y}{X}$$

where X and Y are rational integral functions of the nth degree in x and y...

Formulated in modern terminology, Hilbert's question is about the study of *bifurcations* of limit cycles in the family $\mathcal{P}_n$ of all polynomial vector fields of degree $\leq n$, parametrized by the space of coefficients. Let us give a formulation of a weak version of Hilbert's problem, known presently as the *existential Hilbert's sixteenth problem*:

For any $n \geq 2$, there exists a number $H(n) < \infty$, such that any vector field of degree $\leq n$ has less than $H(n)$ limit cycles.

Of course we have discarded the case $n = 1$ as trivial: $(H(1) = 0)$. Recall that the problem remains open even for $n = 2$.

As was stressed in Hilbert's formulation, one has to study the bifurcations of limit cycles as the parameter varies. Hence, in order to study this question we now introduce the central concept of these notes: the concept of *limit periodic set*, the organizing center for the bifurcations of limit cycles.

2.1 Organizing centers for bifurcations of limit cycles

2.1.1 Definition of limit periodic sets

Let (X_λ) be a $\mathcal{C}^1$ family of vector fields on a surface M, not necessarily of genus 0.

Definition 10 [R3] *A limit periodic set for (X_λ) is a compact non-empty subset Γ in M, such that there exists a sequence $(\lambda_n)_n \to \lambda_*$ in the parameter space P, and for each λ_n, the vector field X_{λ_n} has a limit cycle γ_{λ_n} with the following property:*
  $\gamma_{\lambda_n} \to \Gamma$ *where $n \to \infty$ in the Hausdorff topology of the space $\mathcal{C}(M)$ of all non-empty compact subsets of M.*

Recall that if M is a metrizable space, then the Hausdorff topology is defined in the set $\mathcal{C}(M)$ of all compact non-empty subsets of M, in the following way: let d be a distance on M, defining its topology. For $A, B \in \mathcal{C}(M)$, let $d_H(A, B) = \text{Sup}_{x \in A, y \in B} \left\{ \text{Inf}_{z \in B} d(x, z) \ , \ \text{Inf}_{z' \in A} d(z', y) \right\}$.

It is not difficult to show that d_H is a distance on $\mathcal{C}(M)$, and that this distance defines a topology on $\mathcal{C}(M)$ independent of the choice of the distance d: the *Hausdorff topology on $\mathcal{C}(M)$*. It is less obvious to show that if (M, d) is a compact metric space, the same is true for $(\mathcal{C}(M), d_H)$ (see [Ba], for instance).

Remark 5 1) *Once the distance d on M is chosen, then the convergence $\gamma_{\lambda_n} \to \Gamma$ is equivalent to the following: for any $\varepsilon > 0$, $\exists n(\varepsilon)$ such that if $n \geq n(\varepsilon)$ then γ_{λ_n} enters the ε-neighborhood of Γ and inversely, Γ enters the ε-neighborhood of γ_{λ_n}.*

2) *Definitions of limit periodic sets have been proposed by many authors, in particular Perko [Per], Françoise and Pugh [FP], and more recently Ilyashenko and Yakovenko [IY2]. These definitions are more restrictive in the sense that* Γ *is supposed to be a limit of a 1-parameter family of limit cycles* $\Gamma = \lim\limits_{\varepsilon \to 0} \gamma_{\lambda(\varepsilon)}$ *for a continuous arc* $\lambda(\varepsilon) :]0,1[\to P$. *I prefer to introduce a definition in terms of a discrete sequence* $(\lambda_n)_n$, *because it is better adapted to the proofs of the topological properties of limit periodic sets. It is clear that the two definitions are not equivalent for* C^∞ *families. Take the 1-parameter family*

$$y\,\frac{\partial}{\partial x} - x\,\frac{\partial}{\partial y} + \Big(\varphi(\varepsilon) - (x^2 + y^2)\Big)\Big(x\,\frac{\partial}{\partial x} + y\,\frac{\partial}{\partial y}\Big),$$

where $\varphi(\varepsilon) = \sin\,(\frac{1}{\varepsilon})e^{-\frac{1}{\varepsilon^2}}$, $\varphi(0) = 0$. *For this family, the origin of* $\mathbb{R}^2$ *is a limit periodic set in the above sense, but it is not in the "continuous" definition. For analytic families, the equivalence of the two definitions is an open question. The answer would be positive if it were true that the* bifurcation diagram of each limit periodic set in analytic families has a topological conic structure. *But this is again an open question, which is fundamental and surely among the most difficult ones in the whole subject.*

First examples of limit periodic sets on a surface S of genus 0:

– **Singular elliptic points:** They appear, for instance, as limit periodic sets in Hopf-Takens bifurcations.

– **Periodic orbits:** A multiple periodic orbit (for instance a double or semi-stable periodic orbit) may bifurcate in several hyperbolic ones.

These two first examples are called regular limit periodic sets. Their bifurcations can be studied using the theory of bifurcations for smooth functions (catastrophe or singularity theory) or analytic geometry (for analytic vector fields, and center point or non-isolated periodic orbits). We will treat them in Chapter 4.

– **Saddle connection:** We have found a saddle connection in the Bogdanov-Takens bifurcation. The study of their bifurcations brings new problems because the return map near such a connection is no longer differentiable. They are studied in Chapter 5, together with more general elementary graphics.

Later, in Chapter 6 we study more degenerate limit periodic sets. Let LC be the union, in the product space $S \times P$, of all limit cycles. This set is as smooth as the family, so it is analytic if the family is analytic. But its closure $\overline{LC}$ is no longer an analytic subset of $S \times P$ in general, except at the regular limit periodic sets. This is why bifurcations of limit cycles cannot be treated entirely by methods of analytic or differentiable geometry.

2.1.2   The structure of limit periodic sets

Lemma 2 *Let Γ be a limit periodic set in a C^1 family X_λ, defined on a surface S of genus 0, for some value $\lambda_* \in P$. Let $\sigma \subset S$ be an interval transversal to X_{λ_*}. Then $\sigma \cap \Gamma$ contains at most one point.*

Proof. Suppose that $\sigma \cap \Gamma$ contains at least two different points a, b. For n large enough, the vector field X_{λ_n} is transversal to σ and γ_{λ_n} cuts σ at two points at least: a_n near a and b_n near b. As $a \neq b$, one also has $a_n \neq b_n$ for large n. But this is impossible by the same arguments as in Lemma 1.1. $\square$

Therefore limit periodic sets have the same basic property as ω-limit sets of individual vector fields: they can be cut at most in one point by any transversal segment. As noted in 1.1.2, this implies the conclusions of Theorem 1.1. Hence we have proved

Theorem 5 *(Poincaré-Bendixson for families of vector fields.) Let X_λ be a C^1 family of vector fields on a compact surface of genus 0. Let Γ be a limit periodic set for this family for the parameter value $\lambda_* \in P$. Assume that all the singular points of X_{λ_*} are isolated. Then Γ is of one of the following three types:*

　　1) Γ is a singular point of X_{λ_*},

　　2) Γ is a periodic orbit,

　　3) Γ contains a subset Σ of singular points and at least one regular orbit. The ω and α limit set of each of these regular orbits is an element of Σ. Moreover, if S is compact and X_λ analytic, then Γ is a graphic.

The result is very similar to the Poincaré-Bendixson theorem for ω-limit sets. Nethertheless, the following differences are worth noting.

– a periodic orbit which is an ω-set must be isolated on one side. By contrast, a non-isolated periodic orbit (for instance a level curve of a Hamiltonian vector field) may be a limit periodic set.

– a graphic which appears as a limit periodic set may be non-monodromic. The simplest example of this phenomenon is the graphic Γ made by a central manifold connection at a saddle point of codimension 1. As noted in Chapter 1, such a singularity unfolds in a 1 dimensional versal unfolding, written locally as $-y \, \dfrac{\partial}{\partial y} + (\lambda + x^2) \, \dfrac{\partial}{\partial x}$. We assume that for $\lambda = 0$, the separatrix which is locally $0x$ for $x > 0$ returns along the separatrix $0x$, $x < 0$, to make the graphic Γ. Then, one hyperbolic attracting limit cycle appears near this connection Γ, for $\lambda < 0$, near 0. Hence, Γ is a limit periodic set of the unfolding and it is not monodromic as shown in Figure 2.1.

We can now give a more accurate classification of possible limit periodic sets for families on compact surfaces of genus 0:

– **Regular limit periodic sets:** Elliptic singular points or periodic orbits. They may be of finite or infinite codimension. In this last case, we always assume that the

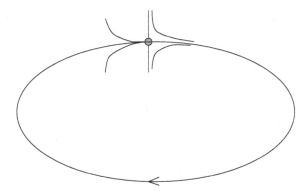

Figure 2.1 Non-monodromic limit periodic set

family is analytic. Infinite codimension will mean that the return map for X_{λ_*} near Γ is equal to the identity (the elliptic point will be called a *center*).

– **Elementary graphics:** Graphics all of whose singular points are elementary. An elementary singular point is an algebraically isolated singular point with at least one real non-zero eigenvalue (they are the irreducible singular points in the sense of the desingularization theory developed in Chapter 3). An elementary point may be hyperbolic, or semi-hyperbolic. In both cases it must be of saddle type (or saddle-node type) in order to be a point of a limit periodic set. Sources or sinks are of course forbidden. If all the singular points are hyperbolic saddles, one says that the graphic is a *hyperbolic graphic*. Elementary graphics may be monodromic or not, isolated (among periodic orbits of X_{λ_*}) or not. In Chapter 5, we will study some of them.

– **Non-elementary graphics:** These are graphics with some non-elementary singular points. We suppose again that all singular points in the graphic are isolated. The simplest example is obtained by connecting the two separatrices of a cuspidal singular point of Bogdanov-Takens type. We call it a cuspidal loop, and we return to it in detail in Chapter 6. In this chapter we will explain how, in some sense, non-elementary graphics can be reduced to elementary ones. We can find many non-elementary graphics even in the family of quadratic vector fields, as we explain at the end of this chapter. Elementary graphics may be monodromic or not, isolated or not among the periodic orbits of X_{λ_*}.

– **Limit periodic sets with non-isolated singular points.**
We will see in Chapter 3 that it is easy to get rid off non-isolated points of an individual analytic vector field. Therefore, the restriction about the singular points in Theorem 1.1 is not too serious. This is not the case for families: one cannot easily replace a given family by a new one such that for each λ, all singular points of X_λ are isolated. The existence of non-isolated singular points is of course a non-generic

phenomenon. But we will see that they appear in a systematic way when using desingularization methods in families (rescaling of variables for instance). Also, they are present in polynomial families. We study these questions in Chapter 6.

A general structure theorem for limit periodic sets (even for analytic families) is not known. In Chapter 6 we will give some partial results in this direction. The *degenerate graphics* will be the simplest examples:

Definition 11 *A degenerate graphic Γ of a vector field X on a surface is a compact, non-empty invariant subset made by a finite number of isolated singular points $\{p_1, \ldots, p_s\}$, regular orbits $\{\gamma_1, \ldots, \gamma_\ell\}$ and arcs of non-isolated singular points $\{r_1, \ldots, r_k\}$, such that the ω and α limit set of each regular orbit is a point in $\{p_1\} \cup \cdots \{p_s\} \cup r_1 \cup \cdots \cup r_k$. Moreover, Γ is the direct image of an S^1-immersion, oriented by increasing time along the regular orbits.*

For instance, all limit periodic sets with non-isolated singular points appearing in the study of quadratic vector fields are degenerate graphics (see [DRR1]).

2.2   The cyclicity property

In this section, we will show that the problem of finding a uniform bound for the number of limit cycles of a given family, for instance Hilbert's 16th problem, can be replaced by a local problem on the number of limit cycles which bifurcate from each limit periodic set.

2.2.1   Definition of cyclicity for limit periodic sets

The following is a precise definition of the number of limit cycles which bifurcate from a limit periodic set:

Definition 12 *Let Γ be a limit periodic set of a C^1 family X_λ, defined at some value $\lambda_* \in P$. Denote by d a distance on S and P and by d_H the induced Hausdorff distance on $\mathcal{C}(S)$. For each $\varepsilon, \delta > 0$ define*

$$N(\delta, \varepsilon) = \text{Sup}\left\{ \text{number of limit cycles } \gamma \text{ of } X_\lambda \mid d_H(\gamma, \Gamma) \leq \varepsilon \right.$$

$$\left. \text{and } d(\lambda, \lambda_\alpha) \leq \delta \right\}.$$

Then the cyclicity of the germ (X_λ, Γ) is given by

$$\mathcal{C}ycl\ (X_\lambda, \Gamma) = \text{Inf}_{\varepsilon, \delta}\ N(\delta, \varepsilon).$$

As indicated in the definition, this bound $\mathcal{C}ycl\ (X_\lambda, \Gamma)$ depends only on the germ X_λ along Γ, i.e., on the unfolding (X_λ, Γ). Of course $\mathcal{C}ycl\ (X_\lambda, \Gamma)$ may be infinite. If it is finite, it represents in a precise way the local bound for the number of limit cycles which bifurcate from Γ *in the given family* X_λ.

A priori, if we change the unfolding (X_λ, Γ) of (X_{λ_*}, Γ), the cyclicity may change. The same finite uniform bound for all the possible unfoldings of (X_{λ_*}, Γ) may exist. In this case, we call it *absolute cyclicity* of (X_{λ_*}, Γ), or simply absolute cyclicity of Γ. It depends only on the germ of the unfolded vector field X_{λ_*} along Γ.

In the next section, we will return to the general question of the relation between the local bounds (finite cyclicity) and the global bounds (as in Hilbert's 16th problem). Here, to conclude this section, we want to emphasize that the computation of the cyclicity may be the crucial step in the process of determining the bifurcation diagrams.

To illustrate this point, we return to the Bogdanov-Takens bifurcation. Let us suppose it is known that any limit periodic set in this family *has cyclicity of less than 1*. Then we can easily deduce Theorem 1.4 from this. It works as follows: fix some value $\mu_0 < 0$ near 0. The effect of increasing the parameter ν is just to make a positive translation on the graph of the return map $P_\nu(x)$ on the interval $[e_{\mu_0}, s_{\mu_0}]$. This comes from the *rotating* property of the vector $X_\lambda(x, y)$. Then, the cyclicity hypothesis implies that a periodic orbit has just one way to appear from a Hopf bifurcation and one way to disappear in a saddle connection bifurcation. It is not possible to create or to annihilate a pair (or more) of periodic orbits. As a consequence, there is no limit cycle for $\nu \notin]\nu_c(\mu_0), \nu_h(\mu_0)[$ and just one limit cycle inside this interval. This suffices to establish the bifurcation diagram and so to prove Theorem 1.4.

2.2.2 The finite cyclicity conjecture.
Local reduction of Hilbert's 16th problem

It is easy to produce $\mathcal{C}^\infty$ families where some limit periodic set has infinite cyclicity. But I am convinced that this is not the case for analytical families. Let us formulate this idea precisely:

– **Finite cyclicity conjecture:** Let Γ be a compact invariant subset of an analytic vector field X, on a surface of genus 0. Then, for any *analytic* unfolding (X_λ, Γ) of (X, Γ), one has $\mathcal{C}ycl\ (X_\lambda, \Gamma) < \infty$.

These notes are essentially devoted to a partial proof of this conjecture. In Chapter 3, we will see that it is valid for 0-parameter families (Dulac's problem), in Chapter 4, that it is valid for regular limit periodic sets and in Chapter 5, that it is valid for unfoldings of generic elementary graphics. Of course, a complete proof of the conjecture remains an open problem. We will show here how it would imply a positive answer to Hilbert's 16th problem.

First, note that a direct consequence of the definition is that the cyclicity is an upper semi-continuous function on the set of all limit periodic sets.

Lemma 3 *Let* $(\lambda_i) \to \lambda_*$ *be a converging sequence in* P. *Suppose that for each* i, Γ_i *is a limit periodic set for the value* λ_i, *such that* $(\Gamma_i) \to \Gamma_*$ *as* $i \to \infty$,

in the Hausdorff sense. Then, Γ_* is a limit periodic set for the value λ_* and
$\mathcal{C}ycl\ (X_\lambda, \Gamma_*) \geq \limsup\limits_{\lambda_i \to \lambda_*} \mathcal{C}ycl\ (X_\lambda, \Gamma_i).$

As a consequence one has the following:

Proposition 1 [R3] *Let X_λ be a C^1 family of vector fields defined on a* compact *surface S of genus 0 with a* compact *set of parameters P. Then, there exists a uniform bound $H((X_\lambda)) < \infty$ for the number of limit cycles of each vector field X_λ (each X_λ has less than $H((X_\lambda))$ limit cycles), if and only if each limit periodic set Γ of (X_λ) has a finite cyclicity in (X_λ).*

Proof. Of course, if such a bound $H((X_\lambda))$ exists, it is trivial that each limit periodic set has a finite cyclicity in (X_λ). Suppose the contrary that $\mathcal{C}ycl\ (X_\lambda, \Gamma) < \infty$ for each limit periodic set in (X_λ) but a finite uniform bound $H((X_\lambda))$ does not exist. Using that $\mathcal{C}(S)$ is compact, this implies that the cyclicity cannot be bounded: one can find a sequence (λ_i) in P, and a limit periodic set Γ_i for each X_{λ_i} such that $\mathcal{C}ycl\ (X_\lambda, \Gamma_i) \to \infty$ for $i \to \infty$. Now, because P and $\mathcal{C}(S)$ are compact spaces, one can find a subsequence λ_{i_j} such that $(\lambda_{i_j}) \to \lambda_*$ and such that $\Gamma_{i_j} \to \Gamma_*$ in the Hausdorff sense. We have that $\mathcal{C}ycl(X_\lambda, \Gamma_{i_j}) \to \infty$ for $j \to \infty$. It follows from Lemma 3 that $\mathcal{C}ycl\ (X_\lambda, \Gamma_*) = \infty$. This contradicts the hypothesis. $\square$

A family defined on a compact surface of genus 0 with a compact set of parameters will be called *a compact family*. The preceding proposition implies that if the finite cyclicity conjecture is true, then any *analytic* compact family (X_λ) would have a uniform bound $H((X_\lambda)) < \infty$ for the number of the limit cycles of each X_λ. In particular, it would imply a positive answer to Hilbert's 16th problem.

Indeed, it is easy to replace the family $\mathcal{P}_n$ of polynomial vector fields of degree $\leq n$ by an analytic compact family on $S^2 \times S^{N-1}$, where S^{N-1} is the unit sphere in $\mathbb{R}^N$ with $N = (n+1)(n+2)$, the space of coefficients. The reason is as follows: for each $\mu \in \mathbb{R}^+$, one has $X_{\mu\lambda} = \mu X_\lambda$ where $\lambda \in \mathbb{R}^N$, so that $X_{\mu\lambda}$ and X_λ are equivalent and one can restrict λ to S^{N-1}. Next, we have seen in 1.1.3.2, how to embed, up to some analytic positive function, a polynomial vector field X of degree n into some analytic vector field $\widetilde{X}$ defined on S^2 (R^2 is identified with $S^2 - \infty$). The same formula embeds the whole family $(X_\lambda) = \mathcal{P}_n$ into an analytic family $(\widetilde{X}_\lambda)$, defined on $S^2 \times S^{N-1}$ (notice that the multiplicative function $(1 + (z\bar{z})^n)^{-1}$ does not depend on λ). The vector field $\widetilde{X}_\lambda \mid$ int S^2 is equivalent to X_λ, and so a bound $\widetilde{H}(n)$ for the family $(\widetilde{X}_\lambda)$ is also a bound for the polynomial family $\mathcal{P}_n(X_\lambda)$.

Remark 6 *The proof of the above proposition is just a compactness argument. It does not give an algorithm to compute $H((X_\lambda))$, even if we had an explicit bound for the cyclicity of every limit periodic set (Note that we do not assume that there exists a uniform bound for the cyclicity of every limit periodic set; this uniformity*

*follows from the proof). So we have a proof of the existence of the bound $H((X_\lambda))$.
It is exactly the same as in the following simple example: suppose that π is the
projection of some simple compact curve $\Gamma \subset \mathbb{R}^2$ on some line δ, and that we know
that any critical point of the projection is a generic fold point; then there exists a
bound $B < \infty$ such that for any $\lambda \in \delta$ the number of points in $\pi^{-1}(\lambda)$ is less than
B. But, depending on the data Γ, π, δ, this bound B can take any finite value. Here,
it is the same: the finite cyclicity conjecture would imply that for each n, Hilbert's
bound $H(n)$ exists, but it does not allow a computation of this bound. We refer
to this problem, "prove that $H(n) < \infty$ exists", as the existential Hilbert's 16th
problem. We hope that this problem is more tractable than the initial one which
can be stated as, for example "prove that $H(2) = 4$".*

2.2.3   A program for solving the existential Hilbert's problem

As stated above, a general conjecture is that any analytic unfolding has a finite
cyclicity. As a direct approach to this conjecture seems somewhat utopian at this
moment, a more reasonable way to address the question of the existence of a
uniform bound $H((X_\lambda))$ for a given analytic family (X_λ) is to follow the program
below:

 – make a list of every limit periodic set which appears in the family (X_λ),

 – show that each such limit periodic set has a finite cyclicity.

In [DRR1], [DRR2] we have followed this program for a compact family
equivalent to the family $\mathcal{P}_2$ of all quadratic vector fields. Recall that Hilbert's
problem is not solved even in this case. In [DRR1], we acomplished the first step of
the program. In the second paper we collected all known results on finite cyclicity
and added some new ones. I will review these two articles briefly and indicate
the progress made since their publication as well as state the principal difficulties
which remain open.

Before taking the first step, we have to chose a "good" compact family equivalent to $\mathcal{P}_2$. In Section 2.2 above, we showed how to obtain one such family in
general, for any $n \geq 2$. Here, for $n = 2$ it is easy to use the specific properties of
quadratic vector fields to obtain a better compact family (with a minimum number of parameters). In fact, we are only interested in vector fields X which have
at least one limit cycle γ. It is well known that this limit cycle bounds a disk D_γ
in $\mathbb{R}^2$, which contains just one singular point, necessarily a focus or a center [Ye].
Hence, translating this singular point to the origin of $\mathbb{R}^2$, and performing a linear
change of coordinates, the vector field X has the following equation:

$$\begin{cases} \dot{x} & = & \alpha x - \beta y + \varepsilon_1\, x^2 + \varepsilon_2\, xy + \varepsilon_3\, y^2 \\ \dot{y} & = & \beta x + \alpha y + \delta_1\, x^2 + \delta_2\, xy + \delta_3\, y^2 \end{cases} \tag{2.1}$$

with $\beta \neq 0$. Of course, one can suppose also that $(\varepsilon_1, \varepsilon_2, \varepsilon_3, \delta_1, \delta_2, \delta_3) \neq 0$. A time
rescaling allows us to suppose that (α, β) belongs to S^1. Using the linear change
of coordinates $(x, y) \to (x, -y)$, we can even suppose that $\beta \geq 0$, i.e., $(\alpha, \beta) \in \mathbb{P}^1$.

Here we have added the non-necessary value $(\alpha, \beta) = (1, 0)$, in order to have a compact domain for (α, β). Next, the transformation $(x, y) \to (\frac{x}{u}, \frac{y}{u})$ transforms the parameter $(\varepsilon_1, \varepsilon_2, \varepsilon_3, \delta_1, \delta_2, \delta_3)$ to the parameter $\frac{1}{u}(\varepsilon_1, \varepsilon_2, \varepsilon_3, \delta_1, \delta_2, \delta_3)$. Hence, it is sufficient to study X_λ for $\lambda \in \mathbb{P}^1 \times S^5$ $\left((\alpha, \beta) \in \mathbb{P}^1, \ (\varepsilon_1, \ldots, \delta_3) \in S^5\right)$.

As was explained in Section 2.3, we can extend (X_λ) to a family $(\widetilde{X}_\lambda)$ in S^2, or better on D^2, blowing up the point at ∞ on S^2. In this way we have obtained an analytic compact family on $S = D^2$ with parameter in $P = \mathbb{P}^1 \times S^5$.

Remark 7 *It is possible to reduce the quadratic part of (2.1) indexed by parameters in S^5 to a normal form indexed by parameters in S^4. Several such reductions are available: Kaypteyn's, Lienard's, Ye's normal forms. Each of them uses a rotation in the parameter space, to eliminate one coefficient of the quadratic part. But this rotation is not unique in general. Hence, the passage to the normal form is not achieved in a continuous way and then does not preserve the notion of neighbors of vector fields. In order to describe the results it is better to remain in the 6-parameter family $(\widetilde{X}_\lambda)$, $\lambda \in \mathbb{P}^1 \times S^5$ (see [DRR1]).*

As we just have to study the limit cycles surrounding the origin, we do not have to take into consideration all the limit periodic sets in the family, but we only have to prove that:

– Any limit periodic set of $\widetilde{X}_\lambda$ *surrounding the origin* has finite cyclicity (such a limit periodic set may be equal to the origin itself, or bound a disk containing the origin in its interior).

Comparing this question to the initial Hilbert's problem for $\mathcal{P}_2$, we have obtained a substantial reduction.

For instance, we do not need to study the following problems:

(1) quadratic perturbation of linear or constant vector fields (by the way, a question equivalent to Hilbert's problem itself!),

(2) finite cyclicity of singular points of nilpotent linear part (for instance, of cuspidal Takens-Bogdanov singular points),

(3) finite cyclicity of singular points with vanishing linear part,

(4) finite cyclicity of degenerate graphics with lines or curves of non-normally hyperbolic singular points,

(5) investigation of the number of zeros of Abelian integral on intervals of periodic solutions (because we look at the existential Hilbert problem, the weak Hilbert's 16th problem as defined by V. Arnold [I3] is not our aim).

This reduction looks a bit mysterious. For instance the quadratic Bogdanov-Takens family $\dot{x} = y$, $\dot{y} = x^2 + \mu + y(\nu \pm x)$ is a subfamily of $\mathcal{P}_2$ and contains limit periodic sets. Of course all limit cycles of this family exist inside our family $\widetilde{X}_\lambda$. The fact is that, when $(\mu, \nu) \to (0, 0)$, the corresponding parameter value λ tends to some λ_0 (after extracting a subsequence), due to the compacity of the parameter space. It happens that the corresponding limit cycle γ_λ converges toward a limit periodic set of $(\widetilde{X}_\lambda)$ which may be the origin, a periodic orbit or a saddle connection, or perhaps a limit periodic set containing a part of the circle at infinity (the origin in the phase space of the Bogdanov-Takens family, which is a limit periodic set of it, has been blown-up in our new family $(\widetilde{X}_\lambda)$, in the precise sense explained in Chapter 6 below).

So let us look at limit periodic sets surrounding the origin. These possible limit periodic sets are:

(a) limit periodic sets with uniquely isolated singular points,
(b) limit periodic sets with some non-isolated singular points.

For the first class, we can apply the Poincaré-Bendixson Theorem 1.2. As a possible limit periodic set we have the origin (and then $\alpha = 0$), periodic orbits or graphics. We will see in Chapter 4 that regular limit periodic sets have finite cyclicity, so we only need to consider graphics. To obtain the list of such possible graphics, one uses the following information, available for quadratic systems:

- a quadratic vector field has at most 4 singular points in $\mathbb{R}^2$, counted with multiplicity,
- a quadratic vector field has at most 6 singular points at infinity (counted with multiplicity), which appear in opposite symmetrical pairs,
- a line in $\mathbb{R}^2$ has at most 2 contact points with a quadratic vector field, or is invariant,
- a polycycle (i.e., a monodromic graphic) with at least 2 singular vertices must contain the straight line segment joining any pair of vertices.

Using these properties, it is not difficult but rather tedious to obtain a list of all possible graphics. To present them in a rational way we introduced some interesting subcategories of graphics in [DRR1]. Figures 2.2, 2.3, 2.4 which present some of them come from [DRR1]:

– **finite graphics** (i.e., graphics contained in $\mathbb{R}^2$). These graphics have less than 3 vertices. They may be elementary (hyperbolic or not) or non-elementary. They may or may not be monodromic. We have 10 such finite graphics. See Figure 2.2.

– **infinite graphics** (i.e., containing a part of the circle at infinity). We have classified them by the number of their vertices at infinity, their total number of vertices, the nature of the vertices (the simplest ones, with a pair of opposite points at infinity are the 'hemicycles', see Fig. 2.3). These graphics form the large majority, 100 out of a total of 121.

Next, we have to consider limit periodic sets with non-isolated singular points. Recall that the general structure of such limit periodic sets is unknown for a general analytic family. But fortunately, they are not so frequent in our family $(\widetilde{X}_\lambda)$. In fact, if a vector field of $(\widetilde{X}_\lambda)$ has non-isolated singular points it is equivalent, up to some linear change of coordinates, to one of the following vector fields $\widetilde{X}$:

(a) $\begin{cases} \dot{x} = (\lambda x - y)(x + 1) \\ \dot{y} = (x + \lambda y)(x + 1) \end{cases}$ $\lambda \in \mathbb{R}$

(b) $\begin{cases} \dot{x} = \lambda x - y + x^2 \\ \dot{y} = x + \lambda y + xy \end{cases}$ $\lambda \in \mathbb{R}$

(c) $\begin{cases} \dot{x} = x(x + 1) \\ \dot{y} = y(x + 1) \end{cases}$

In the first case $\widetilde{X}$ has a line of singular points $\{x = -1\}$. In the second case, the singular set is the circle at infinity. Finally, in the third case we have the union of $\{x = -1\}$ and the circle at infinity as set of singular points.

In each case, it is easy to see that the possible limit periodic sets are all degenerate graphics, according to Definition 2, made by the union of a regular orbit and one or two segments of singular points. Finally, all five possibilities are illustrated in Figure 2.4. Notice that each representation may contain different degenerate graphics.

This achieves the first part of the program. A few of the 121 different limit periodic sets were known to have finite cyclicity at the time we wrote [DRR1]: the finite graphics $(F_1^1), (F_2^1), F(F_3^1), (F_1^2)$ in Figure 2.2 and the hemispheres $(H_1^1), (H_2^1)$ in Figure 2.3. In [DRR2], we add 25 new cases to this list. All of them are elementary graphics as are those known previously, and have a cyclicity less than 2. (For some of them the result was only obtained under generic assumptions.)

In Chapters 4, 5, we introduce methods to treat regular and elementary limit periodic sets. They are the methods used in [DRR2].

Since the publication of [DRR1] some new results have been obtained: in [DER], Dumortier, El Morsalani and Rousseau proved the finite cyclicity of almost all elementary graphics of finite codimension; Mourtada, ElMorsalani and I treated the case of some infinite codimension hyperbolic graphics with 2 vertices at infinity ([EMR]); this was next extended by the two first authors to finite graphics of the same type ([EM]); finally, Zoladek obtained the finite cyclicity for infinite codimension "triangles" ([Z]).

Till now, none of the non-elementary or degenerate graphics have been studied. In Chapter 6, we introduce a method of desingularization for vector field families. We will verify that using this method we can reduce the question in our family $\widetilde{X}_\lambda$ to a problem of finite cyclicity for a *singular* elementary limit periodic set. These singular limit periodic sets are a little more general than those introduced in this chapter; they will be defined in Chapter 6.

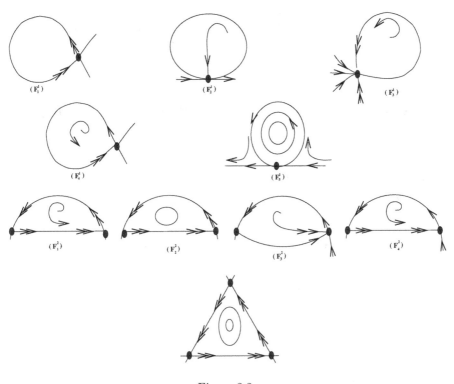

Figure 2.2

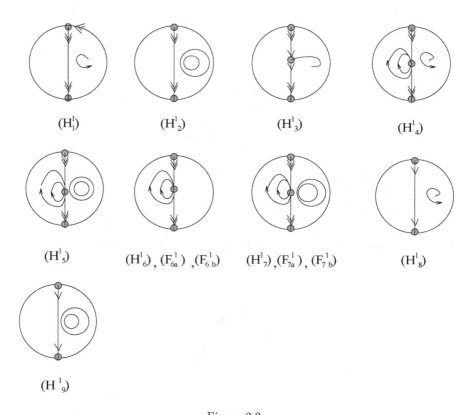

(H^1_1) (H^1_2) (H^1_3) (H^1_4)

(H^1_5) (H^1_6) , (F^1_{6a}) , (F^1_{6b}) (H^1_7) , (F^1_{7a}) , (F^1_{7b}) (H^1_8)

(H^1_9)

Figure 2.3

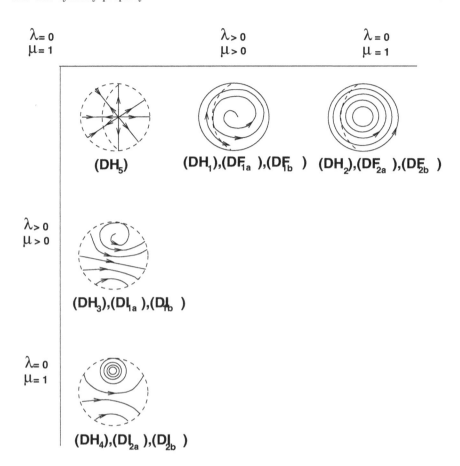

Figure 2.4

Chapter 3
The 0-Parameter Case

As an introduction to the theory of bifurcations, in this chapter we want to consider individual vector fields, i.e., families of vector fields with a 0-dimensional parameter space. We will present two fundamentals tools: the desingularization and the asymptotic expansion of the return map along a limit periodic set. In the particular case of an individual vector field these techniques give the desired final result: the desingularization theorem says that any algebraically isolated singular point may be reduced to a finite number of elementary singularities by a finite sequence of blow-ups. If X is an analytic vector field on S^2, then the return map of any elementary graphic has an isolated fixed point. As a consequence, in this special case there is no accumulation of limit cycles in the phase space. In other words, the cyclicity of each limit periodic set is less than one and any analytic vector field on the sphere has only a finite number of limit cycles.

In the following chapters, we will apply these techniques to families of vector fields. For these, however, they have not been developed as successfully as for individual vector fields. In particular, the main problems which have a solution for individual vector fields remain open for families.

The following text is a survey of the subject which is included for the sake of completeness. We closely follow the texts of F. Dumortier [D1], [D2] concerning desingularization and also texts by Il'yashenko [I2], [I3] and Moussu [Mo] concerning the Dulac problem.

3.1 Blowing up of singularities of vector fields

In this paragraph, a vector field X is studied locally in a neighborhood of a singular point. Hence, we can suppose that the phase space is $\mathbb{R}^2$ and that the singular point is the origin. We suppose that X is $\mathcal{C}^\infty$.

3.1.1 Polar and directional blow-up

Let X be a $\mathcal{C}^\infty$ vector field on $\mathbb{R}^2$, such that $X(0) = 0$.

R. Roussarie, *Bifurcations of Planar Vector Fields and Hilbert's Sixteenth Problem*, Modern Birkhäuser Classics, DOI: 10.1007/978-3-0348-0718-0_3, © Springer Basel 1998

We consider the polar coordinate mapping $\Phi : S^1 \times \mathbb{R} \to \mathbb{R}^2$ given by $\Phi(\theta, r) = (r \cos \theta, r \sin \theta)$. The pull-back $\widehat{X}$, with $\Phi_*(\widehat{X}) = X$, is a $\mathcal{C}^\infty$ vector field on $S^1 \times \mathbb{R}$, called polar blow-up of X.

The smoothness of $\widehat{X}$ is clear from the direct computation of $\widehat{X}$. In fact, let $X = X_1(x,y) \dfrac{\partial}{\partial x} + X_2(x,y) \dfrac{\partial}{\partial y}$ and let $\widehat{X} = \eta_1(\theta, r) \dfrac{\partial}{\partial \theta} + \eta_2(\theta, r) \dfrac{\partial}{\partial r}$. Write $\langle u, v \rangle = u_1 v_1 + u_2 v_2$ for $u = (u_1, u_2)$, $v = (v_1, v_2) \in \mathbb{R}^2$. Since

$$\Phi_*\left(\frac{\partial}{\partial \theta}\right) = x \frac{\partial}{\partial y} - y \frac{\partial}{\partial x} , \quad \Phi_1\left(r \frac{\partial}{\partial r}\right) = x \frac{\partial}{\partial x} + y \frac{\partial}{\partial y} \tag{3.1}$$

we have:

$$\begin{aligned}
\langle \widehat{X}, \frac{\partial}{\partial \theta} \rangle &= r^2 \, \eta_1 = \langle X, \, x \frac{\partial}{\partial y} - y \frac{\partial}{\partial x} \rangle \\[2mm]
\langle \widehat{X}, r \frac{\partial}{\partial r} \rangle &= r^2 \, \eta_2 = \langle X, \, x \frac{\partial}{\partial x} + y \frac{\partial}{\partial y} \rangle.
\end{aligned} \tag{3.2}$$

This gives

$$\begin{aligned}
\eta_1(\theta, r) &= \frac{1}{r^2} \left(-r \sin \theta \, X_1(r \cos \theta, \, r \sin \theta) \right. \\
&\qquad \left. + r \cos \theta \, X_2(r \cos \theta, \, r \sin \theta) \right) \\[3mm]
\eta_2(\theta, r) &= \frac{1}{r^2} \left(+ r \cos \theta X_1(r \cos \theta, \, r \sin \theta) \right. \\
&\qquad \left. + r \sin \theta X_2(r \cos \theta, \, r \sin \theta) \right).
\end{aligned} \tag{3.3}$$

Now, because $X_1(0,0) = X_2(0,0) = 0$, the term r^2 can be factorized in the two parentheses of (3.3), and η_1, η_2 turn out to be $\mathcal{C}^\infty$.

This follows from the Taylor formula with integral remainder. The same idea proves that $\widehat{X}$ is $\mathcal{C}^{k-1}$ if X is $\mathcal{C}^k$.

We verify that $j^k X(0) = 0$ implies that $j^{k-1} \widehat{X}(u) = 0$ for all $u \in S^1 \times \{0\}$, if k is greater than one. This means that the degenerate singularity has been transformed into a whole circle of singularities.

In practice, we simplify the calculations by looking at charts and performing the so-called *"directional"* blow-up,

$$x\text{-direction} : (\bar{x}, \bar{y}) \to (\bar{x}, \bar{y}\bar{x}) \tag{3.4}$$

$$y\text{-direction} : (\bar{x}, \bar{y}) \to (\bar{x}\bar{y}, \bar{y}). \tag{3.5}$$

On $\{x \neq 0\} = \left\{ \theta \neq \dfrac{\pi}{2}, \dfrac{3\pi}{2} \right\}$, (3.4) is the same as a polar blow-up, up to the analytic coordinate change $(\theta, r) \to (r \cos \theta, tg \, \theta)$.

Indeed, $(r\cos\theta, tg\theta.r\cos\theta) = (r\cos\theta, r\sin\theta)$.

The "pull-back" of the directional blow-up maps are hence merely expressions of $\widehat{X}$ in well-chosen coordinate systems. Here the degenerate singularity is transformed into a line of singularities.

After blowing up the singularity to a circle, one can *desingularize* $\widehat{X}$ by considering $\overline{X} = \dfrac{1}{r^k}\,\widehat{X}$, where k is the highest order of zero jets of X at 0, i.e., $j^\ell X(0) = 0$ if $\ell \le k$ and $j^{k+1}\,X(0) \ne 0$.

For the directional blow-up we use $\dfrac{1}{\overline{x}^k}\,\widehat{X}$, resp. $\dfrac{1}{\overline{y}^k}\,\widehat{X}$. These last vector fields are no longer coordinate expressions of $\overline{X}$ but are equal to $\overline{X}$ up to an analytic coordinate change and *multiplication by a positive analytic function*. This positive factor does not constitute any problem since we are only concerned with the orbit structure (phase portrait) of X around the singularity.

Example 1.

Let $X = (x^2 - 2xy)\,\dfrac{\partial}{\partial x} + (y^2 - xy)\,\dfrac{\partial}{\partial y} + 0(\|(x,y)\|^2)$.

Putting

$$c = \cos\theta, \ \ s = \sin\theta,$$

we have:

$$\overline{X} = \frac{1}{r}\,\widehat{X} = 2sc(s - c)\frac{\partial}{\partial\theta} + \tag{3.6}$$
$$(c^3 + s^3 - cs(c + s))r\frac{\partial}{\partial r}.$$

$\overline{X}$ has six singular points for $\theta = 0, \pm\dfrac{\pi}{2}, \pi, \dfrac{\pi}{4}, -\dfrac{3\pi}{4}$, which are all hyperbolic. The phase portraits for $\overline{X}$ near $S^1 \times \{0\}$ and X near the origin are shown in Figure 3.1.

3.1.2 Successive Blow-ups

In the above example, one blow-up was sufficient to determine the topological type of the germ. The reason is that after just one blow-up, all the new singular points are hyperbolic and so have a well determined topological type. Then the different topological types glue up to determine the topological type of the germ X.

It is easy to give examples of vector fields with singularities where one blow-up will not suffice to determine their topological type.

Example 2.

Let $Y_b = y\,\dfrac{\partial}{\partial x} + (x^2 + bxy)\,\dfrac{\partial}{\partial y} + 0(\|(x,y)\|^2)$ (a cusp singularity as defined in Chapter 1). Three steps are required to desingularize it (see below for a precise definition of desingularization) and identify it topologically as a "cusp". See the details of the computation in [T3]. The three steps are represented in Figure 3.2.

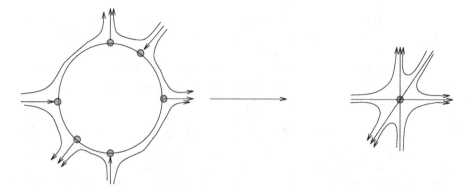

Figure 3.1 A singularity and its polar blow-up.

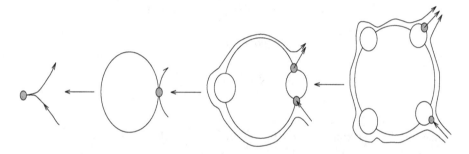

Figure 3.2 Successive blow-ups for the cusp singularity.

The *procedure of successive blow-ups* can be formulated as follows. We use the map

$$\widetilde{\Phi} : \left\{ z \mid \|z\| > \frac{1}{2} \right\} \subset \mathbb{R}^2 \to \mathbb{R}^2 \ : \ z \to z - \frac{z}{\|z\|} \tag{3.7}$$

and then divide out by a power of $(\|z\| - 1)$. To blow up a second time in a point z_0 on the unit circle, we translate it to the origin and again apply $\widetilde{\Phi}$; the second blow-up mapping is therefore, $\Phi_2 = T_{z_0} \circ \widetilde{\Phi}$ where $T_{z_0}(z) = z + z_0$. After a sequence of n blow-ups: $\Phi_1 \circ \cdots \circ \Phi_n$ including the divisions by appropriate powers of r, we find a $\mathcal{C}^\infty$ vector field $\overline{X}^n$ defined on some open domain $U_n \subset \mathbb{R}^2$.

Let $\Gamma_n = (\Phi_1 \circ \cdots \circ \Phi_n)^{-1}(0) \subset U_n$ and denote by A_n the connected component of $\mathbb{R}^2 \backslash \Gamma_n$ with a non-compact closure. We verify that $\partial A_n \subset \Gamma_n$; it is homeomorphic to S^1 and it consists of a finite number of regular $\mathcal{C}^\infty$-arcs meeting transversally at end points. The effect of the divisions is seen as follows: there exists an analytic function $F_n > 0$ on A_n with $\widehat{X}^n = F_n \overline{X}^n$ and $\widehat{X}^n \mid A_n$ is analytically conjugate to $X \mid \mathbb{R}^2 \backslash \{0\}$ by means of $(\Phi_1 \circ \cdots \circ \Phi_n)_{|A_n}$.

Definition 13 *A singular point is called* elementary *if one of the following conditions is fulfilled:*

a) *It is a hyperbolic singularity: the two eigenvalues have non-zero real part.*

b) *It is a non-degenerate semi-hyperbolic singularity: one eigenvalue is nonzero, the other is equal to zero but the infinite jet corresponding to any center manifold is non-zero.*

c) *It is a germ of a line of normally hyperbolic singularities.*

The topological type of an elementary singularity depends only on the sign of the eigenvalues in cases a), c), and also on the principal part of the jet on any center manifold in case b) (see Chapter 1).

Moreover, elementary singularities cannot be simplified by blowing-up: if some of them are blown up only new elementary singularities are produced. So, it is natural to consider them as the final state of the desingularization procedure.

A desingularization theorem for "general" real vector field germs in $\mathbb{R}^2$ was proved by Dumortier [D1]. To express the "generality" of the vector field we need the following definition:

Definition 14 *A vector field X on $\mathbb{R}^2$, with $X(0) = 0$, satisfies a* Lojasiewicz *inequality if there exist $k \in \mathbb{N}$, and $c > 0$ such that*

$$\|X(x)\| \geq c\|x\|^k \text{, for } \forall x \in U,$$

where U is some neighborhood of 0.

This property is not exceptional for germs of vector fields. For instance, a stronger property is:

Definition 15 *A C^∞ vector field has the origin as an* algebraically isolated *singularity if the ideal generated by the components contains a power of the maximal ideal. Notice that this property is equivalent to the similar property for formal series. This property for analytic germs is equivalent to the following topological one: $0 \in \mathbb{C}^2$ is isolated among the zeros of the complexification $\widetilde{X}$ of X.*

It has been proved in [D1] that there exists a subset Σ_∞ of infinite codimension in the space of ∞-jets of vector fields at 0: $J^\infty V$, such that if $j^\infty X(0) \notin \Sigma_\infty$, then X has an algebraically isolated singularity at 0. As a consequence, in any generic family with a finite number of parameters all the singularities are algebraically isolated.

We can now state Dumortier's desingularization theorem.

Theorem 6 *If X is a C^∞ vector field which satisfies a Lojasiewicz inequality, then there exists a finite sequence of blow-ups $\Phi_1 \circ \Phi_2 \cdots \circ \Phi_n$ leading to a vector field $\overline{X}^n$ along ∂A_n whose singularities are all elementary.*

Remark 8 *As $0 \in \mathbb{R}^2$ is an isolated singularity of X, all singularities of $\overline{X}^n$ in some neighborhood V_n of ∂A_n in A_n are on ∂A_n. For instance, the normally hyperbolic singularities (case c) in the above definition) are along smooth arcs of ∂A_n, or may be along all ∂A_n if ∂A_n is smooth (in this case $n = 1$ and $X \simeq (x^2 + y^2)^k \left(x \dfrac{\partial}{\partial x} + y \dfrac{\partial}{\partial y} \right) + O\left(\|(x,y)\|^{2k+1} \right)).*

A consequence of Theorem 1 is that it is always possible to find a finite number of C^∞ invariant lines, each cutting ∂A_n in one point. If this number is non-zero these lines divide a small neighborhood of ∂A_n into a finite number of sectors which after *blowing down*, provide a decomposition of a small neighborhood of the singularity into hyperbolic (or saddle) sectors, elliptic sectors, and parabolic sectors of attracting or expanding type (see Figure 3.3).

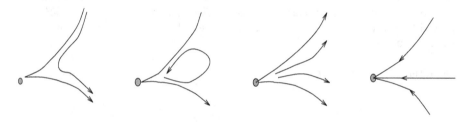

Figure 3.3

The invariant C^∞ lines in the boundary of these sectors blow down to the so-called *characteristic orbits (or lines)*: these orbits tend to the singularity for $t \to +\infty$ or $-\infty$, with a well defined slope. The existence of a characteristic orbit may be read on the vector field $\overline{X}^n$ resulting from the blowing-up procedure. It is equivalent to $\overline{X}^n$ having at least one singular point at a smooth point of ∂A_n. If this is not the case, then one has a well defined return map on a transversal segment to ∂A_n, which blows down to a return map for X on a half segment through the origin. One says that X is of *monodromic type*. In this case, the determination of the topological type of X is more difficult because in general it is not determined by the finite jet of X which determines the desingularization. Hence vector fields with the same algebraically isolated ∞-jet may have different topological types, for instance a vector field of the form

$$j^\infty X(0) = -y \frac{\partial}{\partial x} + x \frac{\partial}{\partial y}.$$

3.1.3 Quasi-homogeneous blow-up and the Newton diagram

Although the method of successive blow-ups is sufficient to study singularities in general, in many cases we can significantly speed up the procedure using *quasi-homogeneous blow-ups*.

Definition 16 *A function $f : \mathbb{R}^n \to \mathbb{R}$ is quasi-homogeneous of type $(\alpha_1, \ldots, \alpha_n) \in \mathbb{N}^n$ and degree k if and only if for any $r \in \mathbb{R}$*

$$f(r^{\alpha_1} x_1, \ldots, r^{\alpha_n} x_n) = r^k \, f(x_1, \ldots, x_n).$$

A vector field $X = X_1 \dfrac{\partial}{\partial x} + X_2 \dfrac{\partial}{\partial y}$ is called quasi-homogeneous of type (α_1, α_2) and degree $k+1$ if X_i is quasi-homogeneous of type (α_1, α_2) and degree $k + \alpha_i$, respectively.

Example: $(ax^2 - 2xy) \dfrac{\partial}{\partial x} + (y^2 - axy) \dfrac{\partial}{\partial y}$ is homogeneous of degree 2: i.e., quasi-homogeneous of type $(1,1)$, and degree 2.

Let us consider an example where the quasi-homogeneous part of lowest degree is determining. For general information on the method we refer to [BM], [Br].

We again consider a cusp singularity Y_b (which requires 3 steps for desingularization by homogeneous blowing-up). An appropriate quasi-homogeneous blow-up for that singularity is:

$$\varphi : S^1 \times \mathbb{R} \to \mathbb{R}^2 \quad (\theta, r) \to (r^2 \cos \theta, \ r^3 \sin \theta).$$

We perform this blow-up in the differential equation for Y_b:

$$Y_b \begin{cases} \dot{x} &= \quad\quad\quad y + 0\big(\|(x,y)\|^2\big) \\ \dot{y} &= x^2 + byx + 0\big(\|(x,y)\|^2\big) \end{cases} \tag{3.8}$$

This gives (writing $\cos \theta = c$, $\sin \theta = s$)

$$\widehat{Y}_b \begin{cases} (3s^2 + 2c^2)\dot{\theta} &= r(2c^3 + 3c^2 - 3) + 0(r^4) \\ (3s^2 + 2c^2)\dot{r} &= r^2 sc(1 + c) + 0(r^4) \end{cases} \tag{3.9}$$

and a desingularized vector field $\overline{Y}_b = \dfrac{1}{r(3s^2 + 2c^2)} \ \widehat{Y}_b$:

$$\overline{Y}_b \begin{cases} \dot{\theta} &= 2c^3 + 3c^2 - 3 + 0(r^3) \\ \dot{r} &= sc(1 + c)r + 0(r^3). \end{cases} \tag{3.10}$$

It is easy to verify that the polynomial $P(c) = 2c^3 + 3c^2 - 3$ has just one (simple) root c_0, $c_0 \in\,]0,1[$, and therefore $\overline{X}_b \mid S^1 \times \{0\}$ has two simple singular points $\theta_0 \in\,]0, \pi/2[$, $-\theta_0$, with $\cos \theta_0 = c_0$.

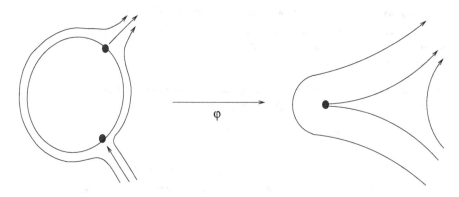

Figure 3.4

At these points, the radial eigenvalue $\pm s_0 c_0 (1 + c_0)$ $(s_0 = \sin \theta_0)$ is non-zero. The phase portraits for $\overline{Y}_b$ and Y_b are given in Figure 3.4.

The determining quasi-homogeneous components can be detected using *Newton's diagram*. The best way to define and also to memorize Newton's diagram is to work with the *dual 1-form* of the given vector field.

For a vector field $X_1 \dfrac{\partial}{\partial x} + X_2 \dfrac{\partial}{\partial y}$, its dual 1-form is $\omega = X_1 \, dy - X_2 \, dx$.

Now take

$$
j^\infty \, \omega(0) = \left(\sum_{\substack{i,j \geq 0 \\ i+j \geq 1}} a_{ij} \, x^i \, y^j \right) dx + \left(\sum_{\substack{i,j \geq 0 \\ i+j \geq 1}} b_{ij} \, x^i \, y^j \right) dy.
$$

The *support* of ω (or X) is defined by

$$
S_\omega = \{(i+1, j) \mid a_{ij} \neq 0\} \cup \{(i, j+1) \mid b_{ij} \neq 0\}.
$$

The *Newton polyhedron* of ω (or X) is the convex hull Γ_ω of the set

$$
P_\omega = \bigcup_{(r,s) \in S_\omega} \{(r,s) + \mathbb{R}_+^2\}
$$

while Newton's diagram of ω (or X) is the union of the compact sides γ_k of Γ_ω. We obtain a quasi-homogeneous component by restricting $(i+1, j)$ and $(i, j+1)$ to some γ_k.

Newton's diagram of the above vector field Y_b has one compact side related to the quasi-homogeneous component $y \dfrac{\partial}{\partial x} + x^2 \dfrac{\partial}{\partial y}$. We have obtained the weights

(2,3) by taking the smallest entire vector orthogonal to the side of Newton's diagram (see Figure 3.5).

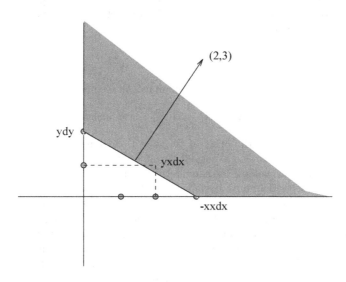

Figure 3.5

3.2   The finiteness result for analytic vector fields on S^2

A preliminary step towards the solution of Hilbert's sixteenth problem is to prove the following finiteness result:

– *A polynomial vector field on $\mathbb{R}^2$ has at most a finite number of limit cycles.*

As we explained in Chapter 1, any polynomial vector field can be extended to an analytic vector field on S^2 and we can ask the previous question for any analytic vector field on S^2.

This question was first studied by Dulac in 1923 [Du2]. He gave an incomplete proof, as several people noticed much later. A correct proof was given for quadratic vector fields by Bamon [Bam]. Very recently, complete proofs of the finiteness result were obtained independently by Ecalle [E] and Il'yashenko [I2]. In this section, we indicate how to reduce this question to the problem of non-accumulation of limit cycles for a polynomial vector field: the so-called *Dulac problem*. In the next section we give some indications of the proof of this last problem in the particular case of hyperbolic polycycles. This case is sufficient to obtain Bamon's result (see [Mo] for details).

Let us consider a polynomial vector field $X = X_1 \dfrac{\partial}{\partial x} + X_2 \dfrac{\partial}{\partial y}$. Such a vector field can have non-isolated zeros, but in this case, the two components X_1 and X_2 have a non-trivial polynomial common factor. Let Q be the greatest common divisor of X_1, X_2. Then

$$X = Q\Big(\overline{X}_1 \frac{\partial}{\partial x} + \overline{X}_2 \frac{\partial}{\partial y}\Big),$$

where $\overline{X}_1$, $\overline{X}_2$ are relatively prime polynomials. As a consequence

– *Any singular point of* $\overline{X} = \overline{X}_1 \dfrac{\partial}{\partial x} + \overline{X}_2 \dfrac{\partial}{\partial y}$ *is algebraically isolated.*

Next, if Γ is a periodic orbit of X, then Γ does not contain any singular points of X. Hence, it is also a periodic orbit of $\overline{X}$. Therefore, X would have a finite number of limit cycles if the same result holds for $\overline{X}$. Hence, we can suppose that our given vector field only has algebraically isolated singular points.

Suppose that such a polynomial vector field has infinitely many limit cycles. The same argument applies for the analytic vector field obtained by extending it to S^2. We again name it X. Using the compacity of the space $\mathcal{C}(S^2)$ of all compact subsets of S^2, we can find a sequence $(\gamma_n)_n$ of limit cycles converging toward some compact invariant subset Γ for X. In the terminology of Chapter 2, Γ is a limit periodic set for the trivial family with 0-parameter, formed of the single vector field X. Looking at Theorem 2.5, we know that Γ is either an elliptic singular point or a periodic orbit, a degenerate singular point or contains at the same time singular points and regular orbits. The first case cannot occur, because the first return map near such a Γ is analytic and cannot have accumulation of fixed points (corresponding to the limit cycles γ_n). For the last case, it was claimed in Theorem 2.5 that Γ must be *a graphic*.

We start by the proof of this claim

Lemma 4 *Let X_λ be an analytic family and Γ a limit periodic set for some value λ_0 which contains singular points and regular orbits of X_{λ_0}. Suppose that each singular point in Γ is algebraically isolated. Then Γ is a graphic.*

Proof. As Γ is compact, it can just contain a finite number of singular points $p_1, \dots, p_k$. To prove that Γ is a graphic, it suffices to prove that Γ also contains only a finite number of regular orbits. Suppose the contrary, that Γ contains infinitely many regular orbits. Then, for at least one of the singular points, say p_1, one has an infinite sequence of regular orbits $(\gamma_n)_{n \in \mathbb{N}}$ such that $\alpha(\gamma_n) = p_1$.

By assumption, p_1 is algebraically isolated and we can apply the desingularization Theorem 1 to it: one has just a finite number of sectors containing p_1. Clearly, the above orbits must belong to elliptic or expanding parabolic sectors. We can find a sub-sequence $(\gamma_{n_i})_{i \in \mathbb{N}}$ in the same sector (which is elliptic or parabolic), and construct a transversal segment σ cutting each γ_{n_i} in one point at least. But this contradicts Lemma 2.2. $\qquad\square$

Remark 9 *The same proof works for limit sets of analytic vector fields. On the other hand, it is possible to construct smooth vector fields which have some limit periodic sets or limit sets with infinitely many regular orbits (and non-algebraically isolated singular points).*

We return now to our vector field X with an accumulation of limit cycles γ_n on some degenerate singular point or some graphic Γ. This set must have a well defined returned map on some half interval σ to Γ ($\sigma \simeq [0,1[$ and $\{0\} = \sigma \cap \Gamma$) on the side of the accumulation. It is called a monodromic *polycycle*.

Now, at each singular point, we can apply the desingularization theorem. The desingularization mapping $\Phi_1 \circ \cdots \circ \Phi_n$ at each point is analytic, so that gluing up local charts defined at each p_i, we can construct an analytic surface $\widetilde{U}$ and a proper map $\Phi : \widetilde{U} \to U$ on some neighborhood of Γ such that Φ is the desingularization map above a neighborhood of each p_i and is an analytic diffeomorphism elsewhere.

The vector field X lifts up to a vector field $\widehat{X}$ on $\widetilde{U}$, which may be desingularized by division by functions defined locally. In this way, we obtain on $\widetilde{U}$ an analytic singular foliation defined by vector fields $\overline{X}_i$. Each $\overline{X}_i$ is defined on an open set U_i, and $\overline{X}_i$, $\overline{X}_j$ differ by a positive analytic function on $U_i \cap U_j$ (we call such a foliation a "local vector field" in Chapter 6). This foliation is oriented and has exactly the same qualitative properties as a vector field. The counter-image $\Phi^{-1}(\Gamma) = \widetilde{\Gamma}$ is an elementary polycycle (each vertex in $\widetilde{\Gamma}$ is elementary). As Φ is an analytic diffeomorphism outside $\widetilde{\Gamma}$, the infinite sequence of limit cycles in U which accumulates on Γ lifts up in an infinite sequence in $\widetilde{U}$ accumulating $\widetilde{\Gamma}$.

Finally, we have reduced the problem to that of proving that such an accumulation is impossible. This is the so-called *Dulac problem*:

– *An elementary polycycle of an analytic foliation cannot be accumulated by limit cycles.*

3.3 The Dulac problem

In this section, we give some indications of the solution of the Dulac problem. As we have said above, this proof is quite recent and I am not sure I understand all its details. Therefore, we will limit ourself to the simpler case of *hyperbolic polycycles*.

In this case, a beautiful proof was given by Il'yashenko in 1985. This proof contains some of the ideas used for the general case and moreover, it can be somewhat extended to general unfoldings of hyperbolic polycycles [EMR]. Here, we will partially follow the survey [Mo] by Moussu.

Let $\mathcal{F}$ be an analytic foliation in a neighborhood U of some hyperbolic polycycle Γ. Let $p_1, \ldots, p_n$ be the vertices labeled in cyclic order. Let σ' be any half-segment transversal to Γ, such that the return map P is defined from $\sigma \to \sigma'$. Here $\sigma \subset \sigma'$ are neighborhoods of the base point $a \in \sigma' \simeq [a, b[$.

At each vertex p_i, one can choose local coordinates (x_i, y_i) such that $0x_i$, Oy_i are local unstable and stable manifolds. More precisely $0x_i^+$, $0y_i^+$ belong to Γ and the trajectories corresponding to the return map near Γ are in the first quadrant.

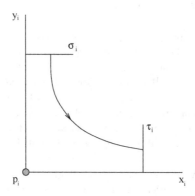

Figure 3.6

Taking transversal segments σ_i, τ_i to $0y_i$, $0x_i$ respectively, we can define a transition map $y_i = D_i(x_i)$ from σ_i^+ to τ_i (σ_i^+ corresponds to $x_i \geq 0$) near each saddle point, and also a regular transition R_i along each side of Γ, from τ_{i-1} to σ_i.

Taking σ, σ' in σ_1^+, for instance, we can write $P(x_1)$ as a composition:

$$P(x_1) = R_n \circ D_n \circ \cdots \circ R_1 \circ D_1(x_1). \tag{3.11}$$

The maps R_i are analytic diffeomorphisms. We want to study the structure of each saddle transition. Let p be a hyperbolic saddle and $D(x)$ its transition map. The structure of D depends strongly on whether the saddle is resonant (with a rational ratio of eigenvalues) or not.

In any case, we will prove in Chapter 5 that there exists a formal series $\widehat{D}(x)$, the so-called *Dulac series of D*:

$$\widehat{D}(x) = \sum_{i=1}^{\infty} x^{\lambda_i} P_i(Lnx),$$

where λ_i is a sequence of positive numbers: $\lambda_1 < \lambda_2 < \cdots < \lambda_1 < \cdots$ tending to infinity, with $\lambda_1 = r = \dfrac{-\mu_2}{\mu_1}$ (the ratio of hyperbolicity of p; μ_2, μ_1 being the eigenvalues), and a sequence of polynomials P_i, with $P_1 = A$ (a positive constant).

This series is asymptotic to $D(x)$ in the following sense. For any s,

$$\mid D(x) - \sum_{i=1}^{s} x^{\lambda_i} P_i(Lnx) \mid = O(x^{\lambda s}). \tag{3.12}$$

Definition 17 *A germ of a map f at $0 \in \mathbb{R}^+$ is called* quasi-regular *if*

(i) *f has a representative on $[0, A[$ which is C^∞ on $]0, A[$.*
(ii) *f is asymptotic to a Dulac series $\hat{f}$*

$$\hat{f}(x) = \sum_{i=1}^{\infty} x^{\lambda_i} \, P_i(Lnx), \ \ with \ 0 < \lambda_1 < \lambda_2 < \cdots a \ sequence$$

of positive coefficients tending to ∞ and P_i a sequence of polynomials.
One says that f is a quasi regular homeomorphism *if f is quasi-regular and if $P_1(x) \equiv A$ (a positive constant).*

It is straightforward to verify that the set of all quasi-regular homeomorphisms is a group $\mathcal{D}$ (for the composition of maps) which contains the group Diff_0 of germs of diffeomorphisms fixing the origin.

As a consequence,

Proposition 2 *The Poincaré map P is quasi-regular.*

Remark 10 *This result is also true for C^∞ vector fields or foliations.*

Now, suppose that $\widehat{P}(x) \not\equiv x$, we have that $P(x) - x$ has also a non-zero Dulac series and therefore $P(x) - x$ is equivalent to some $\alpha x^\lambda \, Ln^k \, x$, for $\alpha \neq 0$, $\lambda > 0$ and $k \in \mathbb{N}$.

But this implies that the equation $\{P(x) - x = 0\}$ has no roots in $]0, X]$, for some $X > 0$, contradicting the assumption on accumulation of limit cycles on Γ, and so, of roots of $\{P(x) - x = 0\}$ on $\{x = 0\}$. Hence, the Dulac series of P is identical to x. We want to prove that this implies that $P(x) \equiv x$. It is precisely this crucial step, $\widehat{P}(x) - x \equiv 0 \Rightarrow P(x) - x \equiv 0$ which seems to be missing in Dulac's paper. This gap was filled by Il'yashenko in [I2] in the hyperbolic case. The idea was to prove a more precise property of *quasi-analyticity* for P:

Definition 18 *Let $f : [0, A[\to \mathbb{R}$.*
One says that f is quasi-analytic *if*

(i) *f is quasi-regular.*
(ii) *The map: $X \to f \circ \exp(-X)$ has a bounded holomorphic extension $F(Z)$ on some domain Ω_b of $\mathbb{C}$, defined by $\Omega_b = \{Z = (X + iY) \in \mathbb{C} \mid X > b(1 + Y^2)^{1/4}\}$, where b is a positive real number.*

A consequence of the Phragmen-Lindelöf theorem (see [Ch]) is that for quasi-analytic functions, the mapping $f \to \hat{f}$ is injective.

Lemma 5 *If f is a quasi-analytic function, such that $\hat{f} \equiv 0$, then $f \equiv 0$.*

Proof. For $b' > 0$ large enough, the image of $\mathbb{C}^+ = \{\text{Real } (Z) \geq 0\}$ by $\varphi : Z \to \varphi(Z) = b'(1+Z)^{1/2} + Z$ with $\varphi(0) = b'$ is contained in Ω_b. Let $F(Z) = f \circ \exp(-Z)$ be as in the definition. The function $G = F \circ \varphi$ is a bounded holomorphic function on $\mathbb{C}^+$, and, as $\hat{f} \equiv 0$, there exist real K, K_n for $\forall n \in \mathbb{N}$ such that

$$| G(Z) | < K \ if \ Z \in \mathbb{C}^+ \text{ and}$$
$$| G(X) | < K_n \exp(-nX) \ if \ X \in \mathbb{R}^+.$$

Now let $G_n(Z) = G(Z).\exp(nZ)$.

We apply the Phragmen-Lindelöf theorem, first to the sectors $\{Y \geq 0, X \geq 0\}$ and $\{Y \leq 0, X \geq 0\}$.

As $| G_n(Z) | \leq K \exp(n \mid Z \mid)$, we have for instance

$$\text{Sup}(| G_n(X + iY) |; \ X \geq 0, \ Y \geq 0)$$

$$\leq \text{Sup}(| G_n(X + iY) |, X \, or \, Y = 0) \leq \text{Sup}\{K, K_n\}.$$

Hence, $| G_n(Z) |$ is bounded on $\mathbb{C}^+$ and we can apply the Phragmen-Lindelöf theorem a second time:

$$\text{Sup}(| G_n(Z) |; Z \in \mathbb{C}^+\} \leq \text{Sup}(| G_n(Z) |; Z \in \partial\mathbb{C}^+\} = K.$$

Using this last inequality for $X \in \mathbb{R}^+$,

$$|G_n(X)| \leq K \Longrightarrow |G(X)| \leq K \exp(-nX), \ \forall n \in \mathbb{N}, \ \forall X \in \mathbb{R}^+.$$

Of course, this implies that $G(X) \equiv 0$. $\square$

Therefore, in order to prove that $P(x) \equiv x$ it suffices to prove that $P(x)$ is quasi-analytic. In the composition $P(x) = R_n \circ D_n \circ \cdots \circ R_1 \circ D_1$ each R_i is real analytic at $x = 0$, and so it is a restriction of a local holomorphic diffeomorphism at $\{z = 0\}$. Clearly, such a function is quasi-analytic. The key point is to prove

Theorem 7 *The transition map D at a hyperbolic saddle singularity is quasi-analytic.*

We postpone for a moment the proof of Theorem 7, in order to finish the proof of the Dulac problem for hyperbolic polycycles.

Let $\varphi(Z) = \exp(-Z)$ and let $\varphi^{-1}(z)$ be the branch of $- \text{Log } (z)$ such that $\varphi^{-1}(1) = 0$.

For each mapping $g(x)$ in the composition $P(x) = R_n \circ D_n \circ \cdots \circ R_1 \circ D_1$ the map $G(z) = \varphi^{-1} \circ g \circ \varphi(Z)$ defines a holomorphic diffeomorphism of some domain Ω_b into another domain $\Omega_{b'}$ (as g is quasi-analytic).

Therefore we can lift up the composition P into a composition of holomorphic diffeomorphisms from domains Ω_{b_i} to $\Omega_{b_{i+1}}$, for $i = 1, \ldots, 2n$. Composing with φ, we obtain that P is quasi-analytic. This completes the proof of the Dulac problem.

Proof of Theorem 7.

Write $g(x) = D(x)$, the transition map. We call $\widetilde{X}$ a complex extension of X in some neighborhood $\widetilde{W}$ of $0 \in \mathbb{C}^2$. Up to some multiplicative factor, the differential equation for $\widetilde{X}$ is

$$\begin{cases} \dot{z} & = & z \\ \dot{w} & = & -r(1 + a(z, w))w, \end{cases} \tag{3.13}$$

$r \in \mathbb{R}^+$ is the hyperbolic ratio, and $z = x + ix'$, $w = y + iy'$ are complex coordinates. We can suppose that $a(z, w)$ is holomorphic in a neighborhood of a polydisk $D \times D$ or radius $(1,1)$ and that $|a| < 1/2$; the trajectories of $\widetilde{X}$ define a holomorphic foliation $\mathcal{F}$, transversal to the projection $\pi(z, w) = z$.

   Any path $c : [0, 1] \to D^+$, starting at $z \in D^+$ and ending at $1 \in D^+$ has a *partial lift* $\bar{c}$ for π, starting at $\bar{z} = (z, 1)$, tangent to $\mathcal{F}$. This means that there exists $\eta > 0$ such that

$$\bar{c} : [0, \eta] \to D^+ \times D, \quad \bar{c}(0) = \bar{z}, \quad \pi \circ \bar{c} = c,$$

and $\bar{c}(t)$ is in a trajectory of $\widetilde{X}$. If $\eta = 1$, we say that $\bar{c}$ is a *lift of c*. If $z = x \in \mathbb{R}^+$ is small enough, then the path $c_x = t \to (1 - t)x + t$, $t \in [0, 1]$ has a lift $\bar{c}_x$ and by definition of g, $\bar{c}_x(1) = (1, g(x))$.

   Let $Z = X + iY$, and let Γ_Z be the composition of the two paths

$$\Gamma_Z^1 : t \to (1 - t)X + iY + t \quad \text{and} \quad \Gamma_Z^2 : t \to 1 + i(1 - t)Y.$$

   If $c_Z = \exp \circ(-\Gamma_Z)$ has a lift $\bar{c}_Z$ we have $(1, G(Z)) = \bar{c}_Z(1)$.

   It is clear that G is holomorphic and bounded in a neighborhood of $\mathbb{R}^+ \subset \mathbb{C}$. It remains to show that this neighborhood contains a domain Ω_b.

   The path c_Z is the composition of the two paths $c_Z^1 = \exp(-\Gamma_Z^1)$ and $c_Z^2 = \exp(-\Gamma_Z^2)$. It is convenient to parametrize these two paths by the flow of the first line of (3.13),

$$c_Z^1 \quad : \quad t \in [0, -\operatorname{Log}|z|] \to ze^t$$

and $$\tag{3.14}$$

$$c_Z^2 \quad : \quad t \in [0, Y] \to e^{-Y}e^{it},$$

where $z = \exp(-Z)$.

   We have to find an inequality $Y \le \varphi(X)$ (for φ of smaller order than X^2) such that one can lift the path c_Z, i.e., lift the path c_Z^1 in a path $\bar{c}_Z^1$ from $m_0 = (z, 1)$ to $m_1 = \left(\dfrac{z}{|z|}, w_1\right)$ and next c_Z^2 in a path $\bar{c}_Z^2$ from m_1 to $m_2 = (1, w_2)$. (See Figure 3.7.)

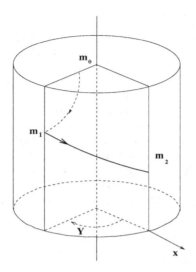

Figure 3.7

To obtain $\bar{c}_Z^1$, we replace the complex time τ in (3.13) by $\tau = t \in [0, -\text{ Log } | z |]$. Using the fact that $| a | \leq 1/2$, we have that the solution of the second line of (3.13) verifies

$$| \omega(0) | e^{\frac{1}{2}rt} \leq | \omega(t) | \leq | \omega(0) | e^{\frac{3}{2} rt}. \tag{3.15}$$

In particular, c_Z^1 can be lifted, and

$$| z |^{1/2r} \leq | \omega_1 | \leq | z |^{3/2r} . \tag{3.16}$$

To obtain $\bar{c}_Z^2$, we replace τ by $\tau = \theta$, $\theta \in [0, Y]$.
The solution $(z(\theta), \omega(\theta))$ verifies $| z(\theta) | \equiv 1$ and

$$\frac{d\omega}{d\theta} = ir(1 + a)\omega(\theta). \tag{3.17}$$

If we write $\omega = \rho e^{i\varphi}$, $\rho \in \mathbb{R}^+$, $\varphi \in \mathbb{R}$, $\omega(\theta) = \rho(\theta)e^{i\varphi(\theta)}$, then (3.17) is equivalent to

$$\left(\frac{d\rho}{d\theta} + i\rho \frac{d\varphi}{d\theta}\right) = ir (1 + a)\rho. \tag{3.18}$$

or

$$\begin{cases} \dfrac{d\varphi}{d\theta} = r + r \text{ Re } (a)\rho \\ \dfrac{d\rho}{d\theta} = - \text{ Im } (a)\rho \end{cases} \tag{3.19}$$

At this point, we need more information on a, because knowing that $|a|$ is bounded would not be sufficient. In fact, using the Dulac form (see Chapter 5), it is possible to choose holomorphic coordinates (z,w) such that $a(z,w) = 0(|z.w|)$. Here, $a(z,w) = a(z(\theta),w(\theta))$ with $|z(\theta)| \equiv 1$, so that

$$|a(z,w)| = 0(\rho). \tag{3.20}$$

Finally, the second line of (3.19) gives $\dfrac{d\rho}{d\theta} = 0(\rho^2)$. That is, there exists $K > 0$ such that

$$\left|\frac{d\rho}{d\theta}\right| \leq K\rho^2. \tag{3.21}$$

By integration, this differential inequality implies

$$|\rho(\theta)| \leq \frac{\rho(0)}{(1 - 3K\,|\,Y\,|\,(\rho(0))^3)^{1/3}}. \tag{3.22}$$

Recall that $\rho(0) = |\omega_1|$.
In order to have $\bar{c}_Z^2 \subset D \times D$, it will be sufficient to have

$$1 - 3K|Y|\,|\omega_1|^3 \geq |\omega_1|^3. \tag{3.23}$$

Taking into account (3.15), it is sufficient to have

$$|Y| \leq \frac{1}{3K}\,(e^{\frac{3}{2}rX} - 1). \tag{3.24}$$

Clearly, the domain defined by (3.24) contains domains Ω_b.

Remark 11 *If $r \notin \mathbb{R}^+$, or if we just use the boundedness of $|a|$ in place of (3.20), we will obtain a linear inequality $|Y| \leq KX$ in place of (3.24). The domain defined in this way does not contain an Ω_b, and we could not apply the Phragmen-Lindelöf idea.*

Chapter 4
Bifurcations of Regular Limit Periodic Sets

In this chapter, (X_λ) will be a smooth or analytic (in Section 3) family of vector fields on a phase space S, with parameter $\lambda \in P$, as in Chapter 1. Periodic orbits and elliptic singular points which are limits of sequences of limit cycles are called *regular limit periodic sets*. The reason for this terminology is that for such a limit periodic set Γ one can define local return maps on transversal segments, which are as smooth as the family itself. The limit cycles near Γ will be given by a smooth equation and the theory of bifurcations of limit cycles from Γ will reduce to the theory of unfoldings of differentiable functions. In fact, we will just need the Preparation Theorem and not the whole Catastrophe Theory to treat finite codimension unfoldings.

Section 3 will be devoted to ∞-codimension analytic unfoldings. In these cases the vector field X_{λ_0} we unfold is of center type; i.e., it admits a whole annulus of closed orbits. If Γ is any of them, the finite cyclicity of (X_λ, Γ) could be deduced from a general theorem by Gabrielov. I prefer to deduce it in a simple way, using the notion of *Bautin Ideal*, which is interesting in itself and may be used also for singular limit periodic sets where methods from analytic geometry are not sufficient, as we will show in the next chapter. Decompositions in the Bautin Ideal are in fact generalizations of the Melnikov asymptotic formulas. Center-type vector fields appear in special analytic families, for instance the polynomial family $\mathcal{P}_n$; they may also result from the use of rescaling formulas, an example of which we will see in this chapter for the Bogdanov-Takens family.

4.1 The return map

4.1.1 Return map for a periodic orbit

Let Γ be a periodic orbit of X_{λ_0}. We choose a point x_0 on Γ and a smooth open interval σ' containing x_0, transversal to X_{λ_0} at any point. We can find an open subinterval σ of σ' containing x_0, such that

51

i) $\bar{\sigma} \subset \sigma'$,

ii) the (first) *return map* $h_{\lambda_0}(u) : \sigma \to \sigma'$ for the flow of X_{λ_0} is defined.

As a consequence of the implicit function theorem, there exists a neighborhood W of λ_0 in P and a map

$$h(u, \lambda) : \sigma \times W \to \sigma'$$

such that for each $\lambda \in W$, $h_\lambda(u) = h(u, \lambda)$ is the first return map of X_λ, from σ to σ', for each $\lambda \in W$, $h_\lambda(u) = h(u, \lambda)$. Here u is a smooth parametrization of σ', with $x_0 = \{u = 0\}$. This map is smooth and analytic if the family (X_λ) is analytic.

Let $\delta_\lambda(u) = \delta(u, \lambda) = h(u, \lambda) - u$ be the *displacement function*; $\delta : \sigma \times W \to \mathbb{R}$. Fixed points of h_λ, which are the roots of $\{\delta_\lambda = 0\}$, correspond to the intersections of periodic orbits of X_λ with σ. In this way, we obtain all closed orbits of X_λ cutting σ, for λ close enough to λ_0. This is a consequence of Lemma 1.1 which implies that each periodic orbit of X_λ cuts σ in at most one point, and of the fact that σ can be chosen as small as necessary. Let us write this result explicitly:

Lemma 6 *For each $\varepsilon > 0$, there exists a neighborhood $\sigma(\varepsilon)$, of x_0 in σ' such that $u \in \sigma(\varepsilon)$ is a root of $\{\delta_\lambda = 0\}$, $\lambda \in W$, if and only if the orbit γ of X_λ through u is a periodic orbit with $d_H(\gamma, \Gamma) \leq \varepsilon$ (see the definition of the Hausdorff distance d_H in Chapter 2).*

Denote by $d(\lambda_1, \lambda_2)$ a distance on P. Let $\varepsilon, \eta > 0$. If $N(\varepsilon, \eta)$ is the number of isolated roots of $\{\delta_\lambda = 0\}$ in $\sigma(\varepsilon) = \{|\, u \,| \leq \varepsilon\}$, for $\{d(\lambda, \lambda_0) < \eta\}$, it follows from Lemma 6 that

$$Cycl \ (X_\lambda, \Gamma) = \mathrm{Inf}_{\substack{\eta \to 0 \\ \varepsilon \to 0}} \ \{N(\varepsilon, \eta),$$

and that the study of $Cycl \ (X_\lambda, \Gamma)$ reduces to the study of the number of roots of the equation $\{\delta_\lambda = 0\}$ near $u = 0$, for λ near λ_0.

4.1.2   Return map near an elliptic point

Recall that an elliptic point x_0 for X_{λ_0} is a singular point with complex eigenvalues. It may be a focus or center-type point (if surrounded by a whole disk of closed orbits). In any case, such a singular point is non-degenerate. Using the implicit function theorem, it is easy to see that the family is smoothly conjugate to a family with $X_\lambda(x_0) \equiv 0$, for (x, λ) near (x_0, λ_0) and with no singular point other than x_0 in a sufficiently small neighborhood of x_0. Moreover, in a neighborhood W_0 of λ_0, the eigenvalues of X_λ at x_0 are equal to $\beta(\lambda) \pm i\alpha(\lambda)$ with $\alpha(\lambda) \neq 0$. From now on, we suppose that such neighborhoods of x_0 and λ_0 are chosen.

Let $\sigma' \simeq [0, b'[$, be a smooth interval embedded in S with the end point 0 at x_0, and transversal to X_{λ_0} at any point $u \neq 0$.

Lemma 7 *Let $b \in]0, b'[$, b' sufficiently small, and $\sigma = [0, b[$. Then there exists a neighborhood $W \subset W_0$ of λ_0 such that the return map for X_λ, $h_\lambda(u)$, is defined* *from σ into σ'. This map, extended by $h_\lambda(0) = 0$ is a smooth function of (u, λ) (analytic if (X_λ) is analytic).*

Proof. Let Ω be a coordinate chart containing $x_0 : \Omega \simeq \mathbb{R}^2$ with coordinates (x, y) and $x_0 = (0, 0)$. Let $\varphi(r, \theta) = (r \cos\theta, r \sin\theta)$ be the polar coordinate map. As we have seen in Chapter 3, there exists a family $(\widehat{X}_\lambda)$ in $(S^1 \times \mathbb{R}^+) \times W$ such that $\varphi_*(\widehat{X}_\lambda) = X_\lambda \mid \Omega \times W$.

Recall that the 'blown-up' family $(\widehat{X}_\lambda)$, for $X_\lambda(x, y) = X_1(x, y, \lambda)\dfrac{\partial}{\partial x} + X_2(x, y, \lambda)\dfrac{\partial}{\partial y}$ in $\Omega \times W$, is equal to

$$\widehat{X}_\lambda = \eta_1(r, \theta, \lambda)\, \frac{\partial}{\partial \theta} + \eta_2(r, \theta, \lambda)\, r\, \frac{\partial}{\partial r},$$

with

$$\eta_1(r, \theta, \lambda) = \frac{1}{r^2}\left(-rs\, X_1(rc, rs, \lambda) + rcX_2(rc, rs, \lambda)\right)$$

and

$$\eta_2(r, \theta, \lambda) = \frac{1}{r^2}\left(rcX_1(rc, rs, \lambda) + rsX_2(rc, rs, \lambda)\right)$$

Here $c = \cos\theta$ and $s = \sin\theta$.

Hence, $(\widehat{X}_\lambda)$ is smooth (or analytic) if (X_λ) is smooth (or analytic).

Suppose the coordinates are chosen so that:

$$j^1\, X_\lambda(0) = \alpha\left(-y\, \frac{\partial}{\partial x} + x\, \frac{\partial}{\partial y}\right) + \beta\left(x\, \frac{\partial}{\partial x} + y\, \frac{\partial}{\partial y}\right).$$

Then, in polar coordinates,

$$J^1\, \widehat{X}_\lambda(\theta, 0) = \alpha\, \frac{\partial}{\partial \theta} + \beta r\, \frac{\partial}{\partial r} \qquad \text{for } \forall \theta \in S^1.$$

This implies that $\widehat{X}_\lambda$ is non-singular for each $(\theta, 0)$ and that the curve $S^1 \times \{0\}$ is a periodic orbit. If $b' > 0$ is small enough, then the interval $\{0\} \times [0, b'[$ is transversal to $\widehat{X}_\lambda$ for all $\lambda \in W$, and one can choose $b \in]0, b'[$ so that the return map $\widehat{h}_\lambda$ of $\widehat{X}_\lambda$ is defined on $]0, b[\times W$.

Of course, $\widehat{h}_\lambda(r)$ is a smooth (resp. analytic) function of (u, λ) if (X_λ) is smooth (resp. analytic). Under the mapping φ, the interval $\{0\} \times [0, b'[$, (resp. $\{0\} \times [0, b[$) is sent to the interval σ' (resp. σ) in the $0x$-axis. As $\varphi_*(\widehat{X}_\lambda) = X_\lambda$, the return map h_λ for X_λ is defined from σ to σ' and is equal to $\widehat{h}_\lambda$. This concludes the proof. $\qquad\square$

The family of vector fields $(\widehat{X}_\lambda)$ is in fact defined in a whole neighborhood of $S^1 \times \{0\}$ in $S^1 \times \mathbb{R}$. It suffices to take $r \in \mathbb{R}$ in the polar coordinate map, and the return map $\widehat{h}_\lambda$ extends in a full neighborhood Σ of $0 \in \mathbb{R}$. This return map is equal to the return map for X_λ, defined on the $0x$-axis for negative values. This means that $h_\lambda(u)$ extends smoothly to a full neighborhood Σ of 0 (such that $\sigma = \Sigma \cap \{x \geq 0\}$). In the same way it is easy to see that the first return of the flow is a well defined smooth map $T_\lambda(u) : \Sigma \to \mathbb{R}$ with $T_\lambda(u).u \leq 0$. Moreover, the following relation is verified in a neighborhood of 0 in Σ:

$$h_\lambda \circ T_\lambda = T_\lambda \circ h_\lambda \quad \text{(see Figure 4.1).}$$

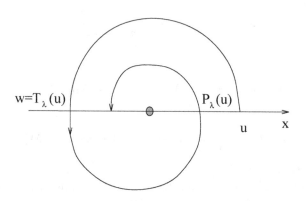

Figure 4.1

It now follows that h_λ on $\{u \leq 0\}$ is locally conjugate to h_λ on $\{u \geq 0\}$ and, if h_λ is a contraction or an expansion for $\{u \geq 0\}$, the same holds for h_λ on $\{u \leq 0\}$. As a direct consequence, we get the following

Lemma 8 *Suppose that $\delta_{\lambda_0}(u) = h_{\lambda_0}(u) - u$ is not flat at $u = 0$ (i.e., $\exists k$ such that $j^k \, \delta_{\lambda_0}(0) \neq 0$). Then $\delta_{\lambda_0}(u)$ has an odd order $2k + 1$:*

$$\delta_{\lambda_0}(u) = (\beta(\lambda_0) - 1)u + o(u), \text{ with } \beta(\lambda_0) \neq 1, \text{ or}$$

$$\delta_{\lambda_0}(u) = +\alpha_{2k+1}(\lambda_0) \, u^{2k+1} + o(u^{2k+1}), \text{ with } \beta(\lambda_0) = 1 \text{ and}$$

$$\alpha_{2k+1}(\lambda_0) \neq 0, k \neq 0.$$

The case $\beta(\lambda_0) \neq 1$ corresponds to a hyperbolic focus. If $\beta(\lambda_0) = 1$ and $\delta_{\lambda_0}(u) = \alpha_{2k+1}(\lambda_0) \, u^{2k+1} + o(u^{2k+1})$ then we say that x_0 is a *weak focus* of X_{λ_0} *of order k.*

4.2 Regular limit periodic sets of finite codimension

4.2.1 Periodic orbit

Let Γ be a periodic orbit for X_{λ_0}, as in 4.1.1, with a transversal interval σ, $\delta_\lambda(u) = h_\lambda(u) - u$ the corresponding displacement function, for $(u, \lambda) \in \sigma \times W (\{u = 0\} = \Gamma \cap \sigma)$.

Definition 19 Γ *is said to be of codimension* $k \geq 0$ *if* $\delta_{\lambda_0}(u)$ *is of order* $k+1$ *at* $u = 0$, *i.e.*,

$$\delta_{\lambda_0}(u) = \alpha_{k+1}\, u^{k+1} + 0(u^{k+1}) \quad \text{with} \quad \alpha_{k+1} \neq 0.$$

Remark 12 *A finite codimension periodic orbit is necessarily a limit cycle. Therefore, Γ is of codimension 0 if and only if Γ is a hyperbolic limit cycle. In this case, one can choose an annulus Ω around Γ and a neighborhood W' of λ_0 in W such that X_λ has an unique (hyperbolic) limit cycle Γ_λ, for all $\lambda \in W'$, X_λ with $\Gamma = \Gamma_{\lambda_0}$. Hence $\mathcal{C}ycl\,(X_\lambda, \Gamma) = 1$, in this case.*

This result of finiteness is easily generalized:

Lemma 9 *Let Γ be a limit cycle of X_{λ_0} of codimension k. Then*

$$\mathcal{C}ycl\,(X_\lambda, \Gamma) \leq k + 1.$$

Proof. As we have seen in Lemma 1, $\mathcal{C}ycl\,(X_\lambda, \Gamma)$ is equal to the number of local roots of the equation $\{\delta_\lambda(u) = 0\}$. But as $\dfrac{\partial^{k+1} \delta_{\lambda_0}}{\partial u^{k+1}}(0) \neq 0$, one can find σ_1 : $0 \in \sigma_1 \subset \sigma$ and a neighborhood $\lambda_0 : W_1 \subset W$ such that $\dfrac{\partial^{k+1} \delta_\lambda}{\partial u^{k+1}}(u) \neq 0$, for $\forall (u, \lambda) \in \sigma_1 \times W_1$. It follows from Rolle's theorem that the function $u \to \delta_\lambda(u)$ has less than $k+1$ roots in σ_1 (for any $\lambda \in W_1$). $\qquad\square$

Remark 13 *If the return map $h_\lambda(u) : \sigma \to \sigma'$ is defined for $\lambda \in W$, then the set of parameter values $\lambda \in W$ for which at least one limit cycle of order k cuts σ is given by the equation*

$$\{\delta_\lambda(u) = \cdots = \delta_\lambda^{(k)}(u) = 0 \;,\; \delta_\lambda^{(k+1)}(u) \neq 0\}.$$

The map which at each $\lambda \in W$ associates $h_\lambda(u) \in \mathcal{C}^\infty(\sigma, \sigma')$ is a smooth map. More generally, if $X_0 \in \chi^\infty(S)$ has a return map $h_{X_0}(u) : \sigma \to \sigma'$, then we can find a neighborhood W of X_0 in $\chi^\infty(S)$ such that each $X \in W$ has a return map $h_X : \sigma \to \sigma'$. The map $X \in \chi^\infty(S) :\to P_X \in \mathcal{C}^\infty(\sigma, \sigma')$ is also smooth (in the sense of differentiable maps between Frechet spaces). It is also easy to prove that the above equations define a codimension k-submanifold $LC_k(\sigma) \subset W$, which is the set of all vector fields in W with a limit cycle of codimension k intersecting σ.

We can call it a singularity as in Chapter 1. This notion is more general than the notion of singularity defined in Chapter 1, which was given by a submanifold in a jet space. Here, $LC_k(\sigma)$ is not defined in terms of the jets of the vector fields but in terms of their return map on σ.

These general singularities are difficult to track in a given family; for instance, the subset LC_k of polynomial vector fields of degree $\leq n$, having at least one limit cycle of codimension k, is an analytic subset of $\mathcal{P}_n$, but we know almost nothing about it. For instance, we do not know if LC_k, for $k \geq 4$ is empty or not in $\mathcal{P}_2$.

It is easy to give a more precise description of unfoldings of codimension k limit cycles. Let σ be a segment transversal to such a limit cycle Γ for the parameter value λ_0. Then $\delta_{\lambda_0}(u) = \alpha_{k+1} u^{k+1} + 0(u^{k+1})$ ($\{u = 0\} = \Gamma \cap \sigma$). It follows from the *Preparation Theorem* that there exist functions $U(u, \lambda)$, with $U(0, \lambda_0) \neq 0$ and $\alpha_0(\lambda), \ldots, \alpha_{k-1}(\lambda)$ in neighborhoods of $(0, \lambda_0)$ and λ_0 respectively such that

$$\delta_\lambda(u) = U(u, \lambda)\left(u^{k+1} + \sum_{j=0}^{k-1} \alpha_j(\lambda)u^j\right). \tag{4.1}$$

If X_λ is analytic, then the functions U, α_j are also analytic [N]. If X_λ is of class $\mathcal{C}^\infty$, then we can find U, α_j of the same class. This is the "$\mathcal{C}^\infty$ Preparation Theorem" of Malgrange [M].

It follows from (4.1) that the equation $\{\delta_\lambda(u) = 0\}$ is equivalent to the polynomial equation

$$u^{k+1} + \sum_{j=1}^{k-1} \alpha_j(\lambda)u^j = 0. \tag{4.2}$$

This equation is factorized through the universal unfolding of the monomial u^{k+1}

$$P_{k+1}(u, \alpha) = u^{k+1} + \sum_{k=1}^{k-1} \alpha_j\, u^j = 0. \tag{4.3}$$

Bifurcation diagrams for the roots of P_{k+1} in terms of the parameter $\alpha = (\alpha_0, \ldots, \alpha_{k-1})$ are well known at least for $k \leq 4$, because they correspond to the first four of the seven "elementary catastrophes" (those reducing to a phase space of dimension 1): the fold for $k = 1$, the cusp for $k = 2$, the swallow tail for $k = 3$ and the butterfly for $k = 4$. We refer to the abundant literature on Catastrophe Theory for a description and also to [D2] for applications to vector fields.

The vector field X_λ is locally equivalent to any vector field with the displacement function:

$$\delta_\lambda^{N\pm}(u) = \pm\left(u^{k+1} + \sum_{j=0}^{k-1} \alpha_j(\lambda)\, u^j\right). \tag{4.4}$$

The first problem is to construct such a vector field family near $\sigma \times \{\lambda_0\}$. It is a trivial exercise in the $\mathcal{C}^\infty$ case and I leave it to the reader:

Lemma 10 *(Lifting Lemma). Let $h_\lambda(u) : \sigma \times W \to \sigma'$ be a C^∞ family of diffeomorphisms of σ into σ'. Then one can find a C^∞ family of vector fields on an annulus U (containing σ'), with parameters in W, having $h_\lambda(u)$ as the first return map.*

Remark 14 *I do not know if such a result is valid for analytic vector field families.*

Clearly, the initial family X_λ is induced through the map $\alpha(\lambda) = \Big(\alpha_0(\lambda), \ldots,$ $\alpha_{k-1}(\lambda)\Big)$ from the versal unfolding $X_\alpha^{k\pm}$ which one can construct, using Lemma 10, for the function $\delta_\alpha^{k\pm}(u) = \pm\Big(u^{k+1} + \sum_{j=0}^{k-1} \alpha_j\, u^j\Big)$. The unfolding $X_\alpha^{k\pm}$ is structurally stable of codimension k, and in any generic l-parameter family, the local unfoldings of a limit cycle are induced by some of the model $X_\alpha^{k\pm}$, $k \leq \ell$.

4.2.2 Elliptic focus

Let us now consider an elliptic focal point or focus x_0 for X_{λ_0}. As above, we can suppose that x_0 is a non-degenerate singular point for any λ belonging to a neighborhood W of λ_0. Let σ' be a transversal segment passing through x_0 $\sigma' \sim [0, b'[$ $(x_0 = \{u = 0\})$ and $\sigma \subset \sigma'$, $\sigma = [0, b[$. We suppose that the return map $h_\lambda(u) : \sigma \times W \to \sigma'$ is given, with $h_\lambda(0) \equiv 0$.

To simplify the study of X_λ and of its return map h_λ, the family is reduced to a normal form. We just recall this notion and refer to [D] for an existence proof:

Up to a C^∞ conjugacy (i.e., a C^∞ coordinate change, depending on the parameter), X_λ is equivalent to

$$
\begin{aligned}
X_\lambda^N &= \Big(f(x^2 + y^2, \lambda) + f_\infty\Big)\Big(-y\,\frac{\partial}{\partial x} + x\,\frac{\partial}{\partial y}\Big) \\
&\quad + \Big(g(x^2 + y^2, \lambda) + g_\infty\Big)\Big(x\,\frac{\partial}{\partial x} + y\,\frac{\partial}{\partial y}\Big),
\end{aligned}
\tag{4.5}
$$

where $f(u, \lambda)$ and $g(u, \lambda)$ are C^∞, $f_\infty(x, y, \lambda)$, $g_\infty(x, \lambda, y)$ are C^∞ and are flat at the origin: $(j^\infty f_\infty(0, \lambda) = j^\infty g_\infty(0, \lambda) = 0)$.

We can write X_λ^N in polar coordinates:

$$
X_\lambda^N = \Big(f(\rho^2, \lambda) + f_\infty(\rho, \theta, \lambda)\Big)\,\frac{\partial}{\partial \theta} + \Big(g(\rho^2, \lambda) + g_\infty(\rho, \theta, \lambda)\Big)\,\rho\,\frac{\partial}{\partial \rho},
\tag{4.6}
$$

with f_∞, g_∞ flat at $\rho = 0$. Of course, $f(0, \lambda) \neq 0$, for any $\lambda \in W$ and we can divide X_λ^N locally along $\{0\} \times W$ by the component on $\dfrac{\partial}{\partial \theta}$.

Hence X_λ is C^∞ equivalent to the family of vector fields

$$
Y_\lambda = \frac{\partial}{\partial \theta} + (G(\rho^2, \lambda) + G_\infty(\rho, \theta, \lambda))\rho\,\frac{\partial}{\partial \rho}.
\tag{4.7}
$$

To obtain the return map on σ (chosen in $\{\theta = 0\}$), we have to integrate the differential equation of Y_λ:

$$\begin{cases} \dot{\theta} & = 1 \\ \dot{\rho} & = (G + G_\infty)\rho. \end{cases} \tag{4.8}$$

We can eliminate the time t, and look for solutions ρ in term of θ. They are solutions of the equation

$$\frac{d\rho}{d\theta} = \Big(G(\rho^2, \lambda) + G_\infty(\rho, \theta, \lambda)\Big)\rho. \tag{4.9}$$

If $\rho(\theta, \lambda)$ is the solution with $\rho(0, \lambda) = u \in \sigma$, then the return map is given by

$$h(u, \lambda) = h_\lambda(u) = \rho(2\pi, \lambda).$$

Now, because (4.9) is a C^∞ equation in ρ^2, up to a flat term, the return map has the following form:

$$h_\lambda(u) = \Big(\overline{h}_\lambda(u^2, \lambda) + h_\infty(u, \lambda)\Big), \tag{4.10}$$

where $\overline{h}_\lambda(u^2, \lambda)$ is C^∞, $h_\infty(u, \lambda)$ is flat at $u = 0$ and $\overline{h}_\lambda = e^{2\pi\beta(\lambda)} + 0(u^2)$. Here $\beta(\lambda) \pm i$ are the eigenvalues of the 1-jet of (4.8). But any flat function can be written as a C^∞ function of u^2 for $u \geq 0$, so that we can include the term h_∞ in $\overline{h}$,

$$h_\lambda(u) = u\,\overline{h}_\lambda(u^2, \lambda), \tag{4.11}$$

giving

$$\delta_\lambda(u) = u\,\overline{\delta}(u^2, \lambda), \tag{4.12}$$

with $\overline{\delta}_\lambda$ C^∞ in u^2 and λ, $\overline{\delta}(u^2, \lambda) = (e^{2\pi\beta(\lambda)} - 1) + 0(u^2)$. Let us suppose that x_0 is a weak focus of order k for X_{λ_0}. This means that

$$\overline{\delta}_{\lambda_0}(u^2) = \overline{\alpha}_k\,u^{2k} + 0(u^{2k}) \quad \text{with } \overline{\alpha}_k \neq 0.$$

In this case, we have trivially:

Lemma 11 *If x_0 is a weak focus of order k, then $Cycl\,(X_\lambda, \{x_0\}) \leq k$.*

Proof. The equation for limit cycles near the origin is $\{\overline{\delta}_\lambda(u^2) = 0\}$ and 0 is a zero of order k of $\overline{\delta}_\lambda$. Applying Rolle's theorem to this function k times gives the result: $\overline{\delta}_\lambda(u^2)$ has less than k zeros on $[0, U]$ for some $U > 0$ and for $\lambda \in W$. $\square$

Of course, as for periodic orbits, one can obtain a precise description for the bifurcation diagram. Applying the Preparation Theorem to the C^∞ function $\overline{\delta}$, we have:

$$\overline{\delta}_\lambda(u) = U(u, \lambda)\Big[u^{2k} + \sum_{j=0}^{k-1} \alpha_j(\lambda)\,u^{2j}\Big], \tag{4.13}$$

$U(0, \lambda_0) \neq 0$ and $\alpha_j(0) = 0$.

Notice that is not possible to eliminate the term in $u^{2(k-1)}$ by a translation in u because we have to preserve $\{u = 0\}$ which corresponds to a singular point of the vector field.

$\overline{\delta}_\lambda$ is then factorized, up to a unity U, through the versal function

$$\overline{\delta}_\alpha(u) = u^{2k} + \sum_{j=1}^{k-1} \alpha_j \, u^{2j}. \qquad (4.14)$$

It is clear that the zeros of $\overline{\delta}_\alpha$ correspond to the limit cycles of the polynomial family of vector fields

$$X_\alpha^{N\pm} = \frac{\partial}{\partial \theta} \pm \left(\sum_{j=0}^{k-1} \alpha_j \, \rho^{2j} + \rho^{2k} \right) \rho \, \frac{\partial}{\partial \rho} \qquad (4.15)$$

($\pm$: sign of $\overline{\alpha_k}$).

Remark 15 *It is in general not possible to eliminate the unity $U(u, \lambda)$ by a conjugacy. Therefore, if $X_{\alpha(\lambda)}^{N\pm}$ is equivalent to X_λ for each λ, it is not possible in general to construct a $(\mathcal{C}^0, \mathcal{C}^0)$ equivalence for $k \geq 4$. We can construct a topological obstruction to this! (See [R1]).*

We have finally obtained that $X_\alpha^{N\pm}$ is the versal unfolding of the germ X_{λ_0}, for the $(\mathcal{C}^0$-fibre, $\mathcal{C}^\infty)$ equivalence relation. This means that we can find a $\mathcal{C}^\infty$ map $\alpha(\lambda)$ such that X_λ is topologically equivalent to $X_{\alpha(\lambda)}^{N\pm}$, for each value of λ. If X_λ is analytic, then the map $\alpha(\lambda)$ is also analytic.

This result was proved for $k = 1$ by Hopf and extended for any $k \geq 2$ by F. Takens [T1]. In fact, Takens proved a somewhat stronger result: he obtained a smooth map $H(x, y, \lambda) : \mathbb{R}^2 \times W \to \mathbb{R}^2 \times W$ above the map $\alpha(\lambda)$ which brings the limit cycles of X_λ into the limit cycles of the model $X_\alpha^{N\pm}$. We, therefore, call the above unfolding a degenerate Hopf unfolding, or a Hopf-Takens unfolding.

4.3 Regular limit periodic set of infinite codimension

We now restrict ourselves to an analytic family X_λ, and we suppose that for some value $\lambda_0 \in P$, there exists an interval σ' such that each orbit cutting it is periodic. We say that X_{λ_0} is of *center-type*. Of course, we may suppose that σ' is a maximal interval with this property and that each end of σ' belongs to some singular limit periodic set. This limit periodic set may be reduced to a single point, a center (hence the name, center-type). The simplest case is when this center is elliptic: the 1-jet of X_{λ_0} at this point is conjugate to a rotation, and this limit point is then a regular limit periodic set like the other orbits through the points of σ'. We will consider such a singular point in this section. The end point of σ' may belong to a more complicated limit periodic set: a single non-elliptic point or a limit periodic set with singular points and regular orbits. We will study this possibility in the next chapter.

If $e \in \partial \sigma'$ is an elliptic point, we will suppose as above that e is a non-degenerate singular point for any value of λ in a neighborhood W_0 of λ_0. In this case, σ' will be a half-closed interval with e as end point.

In all cases (σ' half-closed or open), for any $\sigma \subset \sigma'$ ($e \in \sigma$ if σ' is half-closed), such that $\bar{\sigma} \subset \sigma'$, one can find a neighborhood W of λ_0 in W_0 such that the return map $h_\lambda(u) : (u, \lambda) \in \sigma \times W \to \sigma'$ and the displacement function $\delta_\lambda(u) = h_\lambda(u) - u : \sigma \times W \to R$ are analytic. The condition that X_{λ_0} is of center-type is equivalent to $\delta_{\lambda_0}(u) \equiv 0$. We want to study the cyclicity and bifurcation properties of the germ of X_λ along $\{\lambda_0\} \times \gamma_{u_0}$ when γ_{u_0} is the orbit of X_{λ_0} through $u_0 \in \sigma$ (including the case $u_0 = e$). For this, we introduce in Section 3.1 an ideal of the ring of germs of analytic functions in λ at λ_0: the *Bautin Ideal* $\mathcal{I}$. We will see that the difference function δ_λ may be divided locally in this ideal. We will give also some other properties of $\mathcal{I}$ and of the division of δ_λ which allow us to estimate the cyclicity of (X_λ, γ_u) (Section 3.3). Finally, we will see that the Melnikov asymptotic formula is a special case of division in the ideal and that inversely Melnikov functions may be used to compute the division. We will finish this section with applications of the Bautin Ideal to quadratic vector fields.

4.3.1   The Bautin Ideal

For any $u_0 \in \sigma$, we can expand the analytic function $\delta(u, \lambda)$ in series of $u - u_0$:

$$\delta(u, \lambda) = \sum_{i=0}^{\infty} a_i(\lambda, u_0)(u - u_0)^i.$$

The functions $a_i(\lambda, u_0)$, are analytic in $W \times \sigma$. To simplify notation we will not indicate the dependence on u_0 and will simply write $a_i(\lambda)$ for $a_i(\lambda, u_0)$. Let $\tilde{f}$ denote the germ at λ_0 of an analytic function $f(\lambda)$ defined in a neighborhood of λ_0 ($\tilde{f} \in \mathcal{O}$, the ring of analytic germs at $\lambda_0 \in P$). We consider the ideal $\mathcal{I}^{u_0} \subset \mathcal{O}$, generated by the germs $\tilde{a}_i : \mathcal{I}^{u_0} = \mathcal{I}\{\tilde{a}_i\}_i$.

This ideal is Noetherian and so is generated by a finite number of germs $\tilde{a}_i$:

$$\mathcal{I}^{u_0} = \mathcal{I}\{\tilde{a}_0, \ldots, \tilde{a}_N\}.$$

The functions a_i and the number N depend on u_0.

We have the following *division property*:

Proposition 3 *There exists a constant $R > 0$ such that for all $u_0 \in \sigma$, there exist a neighborhood of λ_0 : $W_{u_0} \subset W$ and analytic functions $h_0(u, \lambda), \ldots, h_N(u, \lambda)$ defined on $([u_0 - R, \, u_0 + R] \cap \sigma) \times W_{u_0}$ and on this domain*

$$\delta(u, \lambda) = \sum_{i=0}^{N} a_i(\lambda) \, h_i(u, \lambda). \tag{4.16}$$

Moreover, $h_i(u, \lambda) = (u - u_0)^i (1 + 0(u - u_0))$.

Remark 16 *Recall that $a_i(\lambda)$ and also N may depend on u_0. But the above constant R is independent of u_0.*

Proof. We suppose that W is compact. Let K be the union of all trajectories of X_λ between points in $\bar\sigma$ and the first return on σ', for $\forall\lambda \in W$; K is a compact subset of $S \times P$. We can extend the real analytic family of vector fields (X_λ) to a holomorphic family of vector fields $(\widehat{X}_{\widehat\lambda})$ defined for $(\hat x, \hat\lambda)$ in some neighborhood of K in the complexification $\widehat S \times \widehat P$ of $S \times P$. For this holomorphic family, we can choose sections $\hat\sigma$, $\hat\sigma' \subset \widehat S$, diffeomorphic to disks, such that $\bar{\hat\sigma}. \subset \text{int } \hat\sigma'$, and $\hat\sigma \cap S = \sigma$, $\hat\sigma' \cap S = \sigma'$. We can also choose some compact extension $\widehat W$ of W in $\widehat P (W = \widehat W \cap P)$, such that $\widehat{X}_{\widehat\lambda}$ has a holonomy map $\hat h(\hat u, \hat\lambda)$: $\hat\sigma \times \widehat W \to \hat\sigma'$. Let $\hat\delta(\hat u, \hat\lambda) = \hat h(\hat u, \hat\lambda) - \hat u$. Then for all $\hat u_0 \in \hat\sigma$, we have a series expansion,

$$\hat\delta(\hat u, \hat\lambda) = \sum_{i=0}^{\infty} \hat a_i(\hat\lambda, \hat u_0)(\hat u - \hat u_0)^i. \tag{4.17}$$

The functions $\hat a_i(\hat\lambda, \hat u_0)$ are holomorphic on $\widehat W \times \bar{\hat\sigma}$ and extend the real function $a_i(\lambda, u_0)$. As $\bar{\hat\sigma} \times \widehat W$ is compact, the expansion (1) has a radius of convergence greater than some $2R > 0$, where R is independent from $\hat u_0$ and $\hat\lambda$. There also exists a constant $M > 0$ such that

$$\mid \hat a_i(\hat\lambda, \hat u_0) \mid \leq M(2R)^{-i} \quad , \text{ for } \quad \forall\, i \in \mathbb{N} \tag{4.18}$$

We want to find holomorphic functions $\hat h_i(\hat u, \hat\lambda)$, $i = 0, \ldots, N$ defined on $D_R(\hat u_0) \times \widehat W_{\hat u_0}$ (where $D_R(\widehat u_0) = \{u \in \hat\sigma \mid \mid \hat u - \hat u_0 \mid \leq R\}$ and $\widehat W_{\hat u_0}$ is some neighborhood of $\hat\lambda_0 = \lambda_0$ in $\widehat W$), such that

$$\hat\delta(\hat u, \hat\lambda) = \sum_{i=0}^{N} \hat a_i(\hat\lambda)\, \hat h_i(\hat u, \hat\lambda). \tag{4.19}$$

The formula (4.16) follows, if we note that for

$$(x, \lambda) \in [u_0 - R, u_0 + R] \cap \sigma \times W_{u_0} \quad , \quad W_{u_0} = \widehat W_{u_0} \cap P$$

we have:

$$\delta(u, \lambda) = \sum_{i=0}^{N} a_i(\lambda) Re[\hat h_i(u, \lambda)].$$

It suffices to take $h_i(u, \lambda) = Re(\hat h_i(u, \lambda))$. To obtain formula (4.19), we have to use the following theorem in [H] (Theorem 7, page 32):

(D) Let $A_0, \ldots, A_N$ be holomorphic functions on a domain V in $\mathbb{C}^\Lambda$ and let $\lambda_0 \in \text{int } V$. Let $\mathcal{I} = \mathcal{I}(\widetilde A_0, \ldots, \widetilde A_N)$ be the ideal generated by germs of the A_i

at λ_0. Then there exist a polydisk $P \subset \text{int } V$, with center at λ_0, and a constant $K > 0$ such that for any function φ holomorphic on P, such that $\widetilde{\varphi} \in \mathcal{I}$, there exist functions $H_0, \ldots, H_N$, holomorphic on P, such that

$$\varphi = \sum_{i=0}^{N} A_i \, H_i \quad \text{on} \quad P$$

and $\mid H_i \mid_P \leq K \mid \varphi \mid_P$.

(Here $\mid . \mid_P$ is the sup norm for continuous functions on P.)

We can apply this to $V = \widehat{W}$ and $\hat{a}_i = A_i$, $i = 0, \ldots, N$. Let $\widehat{W}_{u_0}$, be the polydisk in (D). For each $j > N$, we can write

$$\hat{a}_j(\hat{\lambda}) = \sum \hat{a}_i(\hat{\lambda}).\hat{h}_{ji}(\hat{\lambda}), \tag{4.20}$$

for holomorphic functions $\hat{h}_{ji}$ on $\widehat{W}_{u_0}$, such that

$$\mid \hat{h}_{ji} \mid_{\widehat{W}_{u_0}} \leq K. \mid \hat{a}_j \mid_{\widehat{W}_{u_0}} . \tag{4.21}$$

We can extend formulas (4.20), (4.21) to any $j \geq 0$ by taking $\hat{h}_{ji} = \delta_{ji}$ for $0 \leq i$, $j \leq \ell$, and by replacing K by Sup $\{1, K\}$ in (4.21). Now, in the double sum

$$\hat{\delta}(\hat{u}, \hat{\lambda}) = \sum_{j=0}^{\infty} \left(\sum_{i=0}^{N} \hat{a}_i(\hat{\lambda})\hat{h}_{ji}(\hat{\lambda}) \right) (\hat{u} - \hat{u}_0)^j, \tag{4.22}$$

we can commute the two summations. This is possible because for all i, j

$$\mid \hat{a}_i(\hat{\lambda}).\hat{h}_{ji}(\hat{\lambda})(\hat{u} - \hat{u}_0)^j \mid \leq MK(2R)^{-j} \mid \hat{u} - \hat{u}_0 \mid^j . \tag{4.23}$$

$\square$

Corollary 1 *The ideal $\mathcal{I}^{u_0}$ is independent of the choice of u_0 in σ.*

Proof. Let any $u_0, u_1 \in \sigma$ such that $\mid u_0 - u_1 \mid < R$.

We can apply the formula (4.16) centered at u_0, near u_1, and expand h_i in series of $u - u_1$.

It follows that if

$$\delta(u, \lambda) = \sum_{i=0}^{\infty} b_j(\lambda)(u - u_1)^j, \tag{4.24}$$

then $\tilde{b}_j \in \mathcal{I}^{u_0}$, thus showing that: $\mathcal{I}^{u_1} \subset \mathcal{I}^{u_0}$.

But this argument is symmetrical, so $\mathcal{I}^{u_0} = \mathcal{I}^{u_1}$ if $\mid u_0 - u_1 \mid < R$. The result follows from the connexity of σ. $\square$

Definition 20 *We will call the ideal* $\mathcal{I} = \mathcal{I}^{u_0}$ *for any* $u_0 \in \sigma$ *"Bautin's Ideal". This is an ideal of* $\mathcal{O}_{\lambda_0}$, *the ring of analytic germs at* λ_0. *It is associated to the germ of the return map of* (X_λ) *along* $\sigma \times \{\lambda_0\}$.

Remark 17 $\mathcal{I} \neq \mathcal{O}_{\lambda_0}$ *if and only if* $\delta(u, \lambda_0) \equiv 0$. *If* $\mathcal{I} = \mathcal{O}_{\lambda_0}$, *the function* $\delta(u, \lambda_0)$ *has a finite multiplicity at each* $u_0 \in \sigma$. *The set of zeros of* $\mathcal{I}$: $Z(\mathcal{I})$ *is the germ at* λ_0 *of parameter values for which* X_λ *has a center-type. Bautin computed this ideal for centers of quadratic vector fields(see [B]), which is why it is called "Bautin's Ideal". We will return to Bautin's result in one of the next sections.*

4.3.2   Properties of the Bautin Ideal

Let $\{\widetilde{\varphi}_1, \ldots, \widetilde{\varphi}_\ell\}$ be a set of generators of the Bautin Ideal $\mathcal{I}$. We can write

$$a_i(\lambda) = \sum_{j=1}^{\ell} \varphi_j(\lambda) h_{ji}(\lambda), \quad \text{for} \ \ i = 0, \ldots, N,$$

on some neighborhood W_{u_0} of λ_0 and analytic factors h_{ji}, where a_i are the coefficients of $\delta(u, \lambda)$ at u_0.

Substituting this relation into (4.16) and factorizing, we see that we can divide δ in the functions $\varphi_1, \ldots, \varphi_\ell$.

Proposition 4 *Let* $\varphi_1, \ldots, \varphi_\ell$ *be a set of analytic functions on* W *whose germs generate* $\mathcal{I}$. *Then, for any* $u_0 \in \sigma$, *there exists a neighborhood of* λ_0, $W_{u_0} \subset W$, *and analytic functions* $h_1(u, \lambda), \ldots, h_\ell(u, \lambda)$ *defined on* $[u_0 - R, u_0 + R] \cap \sigma \times W_{u_0}$ *such that*

$$\delta(u, \lambda) = \sum_{i=1}^{\ell} \varphi_i(\lambda) \, h_i(u, \lambda) \tag{4.25}$$

on this domain.

Remark 18 *Of course, we have lost the control of the order of* h_i *in* $u - u_0$.

Definition 21 *We say that* $\{\widetilde{\varphi}_1, \ldots, \widetilde{\varphi}_\ell\}$ *is a minimal set of generators for* $\mathcal{I}$ *if* $\{\widetilde{\varphi}_1, \ldots, \widetilde{\varphi}_\ell\}$ *is a basis of the vector space* $\mathcal{I}/\mathcal{MI}$, *where* $\mathcal{M}$ *is the maximal ideal of* $\mathcal{O}_{\lambda_0}$. *We will call the number of generators of any minimal system* $\ell(\mathcal{I}) = \dim_\mathbb{R} \mathcal{I}/\mathcal{MI}$, *the dimension of* $\mathcal{I}$.

Using Nakayama's lemma, it is possible to extract a minimal set of generators from any set of generators, for instance from the set $\{\widetilde{a}_0, \ldots, \widetilde{a}_N\}$ *of the first coefficients at some point* u_0.

Lemma 12 *Let* $\{\widetilde{\varphi}_1, \ldots, \widetilde{\varphi}_\ell\}$ *be a minimal set of generators and* $\widetilde{f} \in \mathcal{I}$. *Let* $\widetilde{f} = \sum_{i=1}^{\ell} \widetilde{\varphi}_i \, \widetilde{h}_i$ *be a decomposition of* $\widetilde{f}$ *in this set. Then the vector* $(h_i(0))_{i=1,\ldots,\ell}$ *depends only on* $\widetilde{f}$ *and* $\{\widetilde{\varphi}_1, \ldots, \widetilde{\varphi}_\ell\}$ *(note that the decomposition of* $\widetilde{f}$ *is not necessarily unique).*

Proof. $\tilde{f} = \sum_{i=1}^{\ell} h_i(0)\tilde{\varphi}_i$ mod $\mathcal{IM}$ so that $(h_1(0),\dots,h_\ell(0))$ is the vector of $\tilde{f}$-components for $\tilde{f}$ in the basis $\{\tilde{\varphi}_1,\dots,\tilde{\varphi}_\ell\}$ of $\mathcal{I}/\mathcal{IM}$, and is uniquely defined.

$\square$

Lemma 13 *Let $\{\tilde{\varphi}_i\}_i$ and $\{\tilde{\psi}_j\}_j$, $i, j = 1,\dots,\ell(\mathcal{I})$, be two minimal sets of generators. Then there exists a matrix $\{\tilde{H}_{ij}\}$ with coefficients in $\mathcal{O}_{\lambda_0}$ such that $\tilde{\varphi}_i = \sum_{j=1}^{\ell} \tilde{H}_{ij}\, \tilde{\psi}_j$, and the matrix $\{H_{ij}(\lambda_0)\}_{i,j}$ is invertible.*

Proof. Let φ, ψ be the vectors of germs $\{\tilde{\varphi}_i\}$, $\{\tilde{\psi}_j\}$. As these vectors are systems of generators of the same ideal $\mathcal{I}$, there exist matrices of germs H, L such that

$$\varphi = H\psi \quad \text{and} \quad \psi = L\varphi.$$

It follows that

$$\varphi = HL\varphi. \tag{4.26}$$

Then, as a consequence of Lemma 8, $H \circ L(\lambda_0) = H(\lambda_0) \circ L(\lambda_0) = \text{Id}$ and the matrix $H(\lambda_0)$ is invertible.

$\square$

Proposition 5 *Let $\{\tilde{\varphi}_1,\dots,\tilde{\varphi}_\ell\}$ be a minimal system of generators for $\mathcal{I}$. Let $\delta(u,\lambda) = \sum_{i=1}^{\ell} \varphi_i(\lambda)\, h_i(u,\lambda)$ be a division formula, as in Proposition 2, at some point $u_0 \in \sigma$.*

Then the functions $h_i(u) = h_i(u,\lambda_0)$ are independent of u_0 and so globally defined on σ. Moreover, they are $\mathbb{R}$-independent.

Proof. The first part of the conclusion is a consequence of Lemma 12. It suffices to prove the independence of the germs h_i at *some point* $u_0 \in \sigma$, and it suffices to prove this for the factors associated to *some* minimal system of generators $\varphi = \{\tilde{\varphi}_1,\dots,\tilde{\varphi}_\ell\}$. In fact, if $\psi = \{\tilde{\psi}_1,\dots,\tilde{\psi}_\ell\}$ is another minimal system of generators, then by Lemma 8 there exists a matrix of germs H such that $\varphi = H\psi$ and $H(0)$ is invertible. If $h = (\tilde{h}_1,\dots,\tilde{h}_\ell)$ are factors for φ, then

$$\delta = \sum_{i=1}^{\ell} h_i\, \varphi_i = \langle h, \varphi \rangle.$$

But $\langle h, \varphi \rangle = \langle h, H\psi \rangle = \langle\, {}^t\!Hh, \psi \rangle$ for where ${}^t\!H$ is the transposed matrix, so that $h' = {}^t\!Hh$ is a system of factors for ψ and as ${}^t\!H(0)$ is invertible, these factors of h' are $\mathbb{R}$-independent germs at u_0, if this is the case for h.

It therefore suffices to prove the result for a minimal set of generators which is extracted from the system of generators of the coefficients at some $u_0 : \tilde{a}_0, \ldots, \tilde{a}_N$.

Proposition 1 gives a division

$$\delta = \sum_{i=1}^{N} a_i \, h_i,$$

and $H_i(u) = h_i(u, \lambda_0) \simeq (u - u_0)^i$. This last condition implies that the germs h_i, $i = 0, \ldots, N$ are independent at u_0. Unfortunately, the system $\{\tilde{a}_0, \ldots, \tilde{a}_N\}$ is not minimal in general. We are going to extract a minimal system from it by a finite number of steps such that at each step we have a system of generators $\{\tilde{\varphi}_1, \ldots, \tilde{\varphi}_k\}$ obtained from the last system by dropping one term, and such that the associated factors $H_1, \ldots, H_k$ are R-independent. It suffices to prove the recurrence step because after $N - \ell$ steps we must arrive at a minimal set of generators.

Suppose that

$$\delta = \sum_{i=1}^{k} \varphi_i \, h_i, \qquad (k > \ell), \qquad (4.27)$$

with $H_1(x), \ldots, H_k(x)$ R-independent, but such that $\{\tilde{\varphi}_1, \ldots, \tilde{\varphi}_k\}$ is not minimal. This means that one of the $\tilde{\varphi}_i$, say $\tilde{\varphi}_1$, depends on the others mod $\mathcal{IM}$,

$$\tilde{\varphi}_1 = \sum_{j \geq 2} \tilde{s}_j \, \tilde{\varphi}_j \quad \text{mod} \;\; \mathcal{MI}.$$

But this means that there exist $\tilde{m}_1, \ldots, \tilde{m}_l \in \mathcal{M}$ such that

$$\tilde{\varphi}_1 \;=\; \sum_{j \geq 2} \tilde{s}_j \, \tilde{\varphi}_j + \sum_{i=1}^{\ell} \tilde{m}_i \, \tilde{\varphi}_i$$

$$(1 - \tilde{m}_1)\tilde{\varphi}_1 \;=\; \sum_{j \geq 2} (\tilde{s}_j + \tilde{m}_j) \, \tilde{\varphi}_j$$

$$\tilde{\varphi}_1 \;=\; \sum_{j \geq 2} \tilde{S}_j \, \tilde{\varphi}_j \quad \text{for some germs } \tilde{S}_j.$$

Putting this into (4.27):

$$\delta \;=\; \Big(\sum_{j \geq 2} S_j \, \varphi_j \Big) h_1 + \sum_{j \geq 2} \varphi_j \, h_j$$

$$\delta \;=\; \sum_{j \geq 2} k_j \, \varphi_j \quad \text{with}$$

$$k_j \;=\; h_j + S_j \, h_1.$$

The $K_i(u) = k_i(u, \lambda_0)$ are independent (germs).

Suppose, on the contrary, that there exists a non-trivial relation

$$\sum_{i=2}^{k} \alpha_i \, K_i(u) \equiv 0 \quad (\alpha_2, \ldots, \alpha_k) \in \mathbb{R}^{k-1}.$$

This implies that

$$\Big(\sum_{i=2}^{k} \alpha_i \, S_i(0) \Big) H_1(u) + \sum_{i=2}^{k} \alpha_i \, H_i \equiv 0.$$

But, $\{H_1(u), \ldots, H_k(u)\}$ being an independent system, this implies that $\alpha_2 = \cdots = \alpha_k = 0$. This is impossible. $\qquad\square$

The *factor functions* $H_1(u), \ldots, H_\ell(u)$ associated with any minimal system of generators $\widetilde{\varphi}_1, \ldots, \widetilde{\varphi}_\ell$ are analytic. As they are $\mathbb{R}$-independent, each $H_i \not\equiv 0$ and then has some finite order at each $u_0 \in \sigma$. We now prove that for any u_0, it is possible to choose a minimal system having a strictly increasing order of H_i.

Lemma 14 *Let $u_0 \in \sigma$. Then there exists a minimal system of generators such that*

$$order \ H_1(u_0) < \ order \ H_2(u_0) < \cdots < \ order \ H_\ell(u_0) < \infty$$

$$(order \ f(u_0) = n \iff f(u) = \alpha(u - u_0)^n + o((u - u_0)^n), \quad with \ \alpha \neq 0).$$

Proof. Let $(\widetilde{\varphi}_1, \ldots, \widetilde{\varphi}_\ell)$ be any minimal system of generators. Clearly, we can order it so that

$$order \ H_1(u_0) \leq \cdots \leq \ order \ H_\ell(u_0).$$

We will construct a sequence of minimal sets of generators, $\varphi^1, \ldots, \varphi^\ell$, such that $\varphi^1 = (\widetilde{\varphi}_1, \ldots, \widetilde{\varphi}_\ell)$ and such that for φ^s

$$order \ H_1(u_0) < \cdots \ order \ H_{s-1}(u_0) \leq \cdots \leq \ order \ H_\ell(u_0)$$

for the associate system of factors $h = (h_1, \ldots, h_\ell)$.
We now give the recurrence step (how to pass from φ^s to φ^{s+1}, $s < \ell$):

Let $H_1, \ldots, H_\ell$ be the factors for φ^s.
If order $H_s(u_0) <$ order $H_{s+1}(u_0)$, we take $\varphi^{s+1} = \varphi^s$.
If order $H_s(u_0) =$ order $H_{s+1}(u_0) = \cdots =$ order$H_{s+\sigma}$, we take

$$k_i \ = \ h_i \ \text{for} \ s \leq i \ \text{and} \ i \geq s + \sigma$$

$$\text{and} \ k_i \ = \ h_i - \frac{\alpha_i}{\alpha_s} \, h_s \ \text{for} \ s \leq i < s + \sigma.$$

where $H_j(u) = \alpha_j(u - u_0)^{n_j} + \cdots$

These formulas define an invertible matrix M such that: $k = (k_1, \ldots, k_\ell) = Mh$. We have $\langle \varphi^s, h \rangle = \delta = \langle \varphi^s, M^{-1}k \rangle = \langle\, {}^t M^{-1} \varphi^s, k \rangle$. Let $K_j(u) = k_j(u, \lambda_0)$, $K = (K_1, \ldots, K_\ell)$ is the factor function vector for the minimal set ${}^t M^{-1} \varphi^s$. Moreover, it is clear that up to a reordering of terms, we have:

$$\text{order } K_1(u_0) < \cdots < \text{order } K_{s+1}(u_0) \leq \cdots \leq \text{order } K_\ell(u_0). \qquad \square$$

Definition 22 *We say that a minimal system* $\{\widetilde{\varphi}_1, \ldots, \widetilde{\varphi}_\ell\}$ *such that order* $H_1(u_0) < \cdots < \text{order } H_\ell(u_0)$ *is adapted to the point* $u_0 \in \sigma$.

4.3.3 Finite cyclicity of regular limit periodic sets

Let $\{\widetilde{\varphi}_1, \ldots, \widetilde{\varphi}_\ell\}$ be a minimal system of generators, and $H_1(u), \ldots, H_\ell(u)$ be the associated factor functions. As these functions are analytic and $\mathbb{R}$-independent, at each $u_0 \in \sigma$, some of their finite jets are already $\mathbb{R}$-independent.

Definition 23 *For any* $u_0 \in \sigma$, *we define the index* $s_\delta(u_0)$ *by*

$$s_\delta(u_0) = \text{Inf } \{n \in \mathbb{N} \mid \{j^n\, H_i(u_0)\}_i \text{ is an } \mathbb{R}\text{-independent system}\}.$$

As we remarked above, $s_\delta(u_0) < \infty$. It follows from Lemma 13 that this index is independent of the choice of the minimal system. Clearly, if $\{\widetilde{\varphi}_1, \ldots, \widetilde{\varphi}_\ell\}$ is adapted to u_0, $s_\delta(u_0) = \text{order } H_\ell(u_0)$ (the maximal order at u_0, among the factors H_i). This gives a practical way to compute $s_\delta(u_0)$; by looking for an adapted minimal system of generators (see examples below). It also follows from this that $s_\delta(u_0) \geq \ell - 1$ (note that order $H_1(u_0) = 0$ in general!).

When $e \in \partial\sigma$ is a center singular point we need a slightly different definition. Of course, at a center point we have the following:

Lemma 15 *Let* $\delta(u, \lambda) = \displaystyle\sum_{i=1}^{\infty} a_i(\lambda) u^i$ *be the expansion at the center point* e *(corresponding to* $u = 0$*). Then, for* $\forall p \geq 1$: $\tilde{a}_{2p} \in \mathcal{I}(\tilde{a}_1, \ldots, \tilde{a}_{2p-1})$ *That is the ideal is generated by the coefficients of odd order.*

Proof. This property may be obtained using induction formulas for the coefficients a_i (see [B] for instance).

An easier proof involves noting that this property is independent of the choice of the transversal interval σ, the choice of parametrization and also of multiplication of X_λ by an analytic function $g(x, y, \lambda)$, $g(0, 0, \lambda) \neq 0$. It suffices therefore to prove the result when X_λ is written in normal form up to order $2N + 1 \gg 2p$. In this normal form, and in polar coordinates (r, θ) we have

$$X_\lambda = \frac{\partial}{\partial\theta} + \left(\sum_{i=0}^{N} \beta_i(\lambda) r^{2i} + O(r^{2N+2}) \right) r \frac{\partial}{\partial r},$$

$(O(r^{2N+2})$ being an analytic function in (r, θ, λ)).

A direct integration of the differential equation of X_λ gives

$$\dot{\theta} = 1, \ \dot{r} = r\left(\sum_{i=0}^{N} \beta_i \ r^{zi} + 0(r^{2N}) \right).$$

This implies that

$$\delta(r, \lambda) \ = \ a\left(\sum_{j=0}^{N} b_{2j+1}(\lambda)r^{2j} + O(r^{2N+2}) \right),$$

$$b_1 \ = \ (e^{2\pi\lambda_1} - 1)u, \ldots,$$

where $\{u = x = r\}$ is the parametrization of the $0x$-axis.

When (X_λ) is written in normal form the result is trivial because $a_{2i} \equiv 0$ for $i \leq N$.   □

In the division formula we can write each even a_{2p} as a combination of previous odd coefficients. It follows that

$$\delta(u, \lambda) = u \sum_{j=1}^{\ell} a_{2j+1}(\lambda)h_{2j+1} \ (u, \lambda), \tag{4.28}$$

with

$$h_{2j+1}(u, \lambda) = u^{2j}(1 + 0(u)) \tag{4.29}$$

Now, if we extract a minimal system of generators from the system

$$\{a_{2j+1}\}_{j=1,\ldots,\ell},$$

then each of the corresponding factors has an odd order at $e = 0$. This is also true for the adapted system of generators which we can construct from the initial one, as in Lemma 9. As a consequence, for any minimal system of generators

$$\text{Inf} \ \left\{ n \mid \{j^n \ H_j(0)\}_j \ \text{is } \mathbb{R}\text{-independent} \right\} \ \text{is odd.}$$

Definition 24 *If the above number is equal to $2k + 1$ we define $s_\delta(e)$ to be equal to k.*

Finite cyclicity for regular limit periodic sets is a consequence of the following theorem of Gabrielov [G]:

Theorem 8 *(Gabrielov): Let C be a compact analytic real set and $\pi : C \to \Sigma$ be a proper analytic map of C onto another real analytic set. Then, there exists $K < \infty$ such that the number of connected components of $\pi^{-1}(\lambda)$ is bounded by K for any $\lambda \in C$.*

Here, we take $C = \{(u, \lambda) \mid \delta(u, \lambda) = 0 \text{ and } (u, \lambda) \in \bar{\sigma} \times W\}$ and π is the projection onto the parameter space: $\pi(u, \lambda) = \lambda$.

The limit cycles of X_λ through points of σ correspond to the 0-dimensional connected components of $\pi^{-1}(\lambda)$.

The notion of the Bautin Ideal and the related index $s_\delta(u_0)$ allow us to obtain an explicit bound for the cyclicity.

Theorem 9 *Let (X_λ) be an analytic family as above and let $u_0 \in \sigma$ be a point of a transversal interval for X_{λ_0} (u_0 may be a center boundary point). Let γ_{u_0} be the orbit through X_{λ_0} passing through u_0 ($\gamma_{u_0} = e$ if $u_0 = e$). Then,*

   i) *$Cycl\ (X_\lambda, \gamma_{u_0}) \leq s_\delta(u_0)$,*

   ii) *if the Bautin Ideal $\mathcal{I}$ is regular (i.e., $\mathcal{I} = \{\widetilde{\varphi}_1, \ldots, \widetilde{\varphi}_\ell\}$ with $d\varphi_1(\lambda_0) \wedge \cdots \wedge d\varphi_\ell(\lambda_0) \neq 0$), then $Cycl\ (X_\lambda, \gamma_{\lambda_0}) \geq \ell - 1$.*

Proof.

Point i) As was proved in Lemma 6, it suffices to obtain the bound $s_\delta(u_0)$ for (u, λ) belonging to some compact neighborhood $\sigma_1 \times W_1$ of (u_0, λ_0) in $\sigma \times W$. Consider first the case $u_0 \in \text{int}\ (\sigma)$ (u_0 is not a center). By hypothesis, we know that

$$\delta(u, \lambda) = \sum_{i=1}^{\ell} \varphi_i(\lambda) h_i(u, \lambda)$$

in a neighborhood $\sigma_1 \times W_1$ of (u_0, λ_0), with

$$
\begin{aligned}
t_1 = \text{order } H_1(u_0) < t_2 &= \text{order } H_2(u_0) < \cdots < t_\ell \\
&= \text{order } H_\ell(u_0) = s_\delta(u_0) \\
&(h_i(u) = H_i(u, \lambda_0)).
\end{aligned}
$$

Let $\Sigma = \{\lambda \in W_1 \mid \varphi_1(\lambda) = \cdots = \varphi_\ell(\lambda) = 0\}$ and let $W^i = \{\lambda \in W_1 \mid \mid \varphi_i(\lambda) \mid \geq \mid \varphi_j(\lambda) \mid \text{ for } \forall j \neq i\}$, $i = 1, \ldots, \ell$.

Clearly, the zeros of $\delta(u, \lambda)$ are isolated if and only if $\lambda \in W_1 - \Sigma = W^1 \cup \cdots \cup W^\ell - \Sigma$.

We will show that for each $i = 1, \ldots, \ell$, there exists a subinterval σ^i of σ_1, containing u_0 in its interior and a neighborhood W_1 such that for all $\lambda \in W^i - \Sigma$, the function $\delta(u, \lambda)$ has less than t_i isolated roots on σ^i. This interval will be obtained by a succession of restrictive conditions on σ_1 and on W_1, which will be introduced by the claim "restricting u, λ" without more precision.

Fixing $i = 1, \ldots, \ell$, we will construct a sequence of functions $\delta^1 = \delta, \delta^2, \ldots, \delta^i$, by induction as follows:

Restricting u, λ, we can suppose that $\partial^{t_1} h_1(u, \lambda) \neq 0$ for all (u, λ) (W_1 is supposed compact; we note $\partial^s/\partial u^s = \partial^s$).

Then $\partial^{t_1}\delta/\partial^{t_1}h_1 = \varphi_1 + \varphi_2\partial^{t_1}h_2/\partial^{t_1}h_1 + \cdots$.

Let $\delta^2 = \partial(\partial^{t_1}\delta/\partial^{t_1}h_1) = \varphi_2\,h_2^2 + \cdots + \varphi_\ell\,h_\ell^2$.

Clearly, the new functions h_j^2 so defined, verify

$$h_j^2(u,\lambda_0) = \alpha_j^2\,u^{t_j-t_1-1}\,(1+0(u)),\ \text{with}\ \alpha_j^2 \neq 0,$$

and the function δ^2 is similar to δ^1, with one term less. We can introduce the following induction hypothesis, for $2 \leq j \leq i$:

$$\delta^j(u,\lambda) = \varphi_j(\lambda)h_j^j(u,\lambda) + \cdots + \varphi_\ell(\lambda)h_\ell^j(u,\lambda), \tag{4.30}$$

with

$$h_s^j(u,\lambda_0) = \alpha_s^j\,u^{t_s-t_j-1-1}\,(1+0(u)) \tag{4.31}$$

and an induction step defined by

$$\delta^{j+1} = \partial\Big(\partial^{t_j-t_{j-1}-1}\delta^j/\partial^{t_j-t_{j-1}-1}h_j^j\Big), \tag{4.32}$$

for all $j \leq i-1$.

After the last step we obtain the function

$$\delta^i = \varphi_i\,h_i^i + \cdots + \varphi_\ell\,h_\ell^i$$

such that

$$\partial^{t_i-t_{i-1}-1}\,\delta^i = \varphi_i\,k_i + \cdots + \varphi_\ell\,k_\ell$$

where, restricting u,λ, we have $k_i(u,\lambda) \neq 0$, and:

$$k_j(u,\lambda) = 0(u)\ \text{for}\, j > i. \tag{4.33}$$

Then,

$$\mid \partial^{t_i-t_{i-1}-1}\,\delta^i \mid \geq \mid \varphi_i \mid \left(\mid k_i \mid - \sum_{j=i+1}^{\ell}\mid \frac{\varphi_\ell}{\varphi_i}\mid .\mid k_\ell \mid\right). \tag{4.34}$$

As $\mid \frac{\varphi_\ell}{\varphi_i}\mid \leq 1$ on W^i and (4.33), restricting u,λ, we can suppose that $\mid k_i \mid -\sum_{j=i+1}^{\ell}\mid \frac{\varphi_\ell}{\varphi_i}\mid .\mid k_\ell \mid \geq b > 0$ for some constant b. But on $W^i - \Sigma$, $\mid \varphi_i \mid > 0$ and so $\partial^{t_i-t_{i-1}-1}\,\delta^i\,(u,\lambda) \neq 0$ for $\forall(u,\lambda) \in (W^i - \Sigma) \times \sigma^i$.

This last function was obtained from δ by a sequence of $t_1 + 1 + (t_2 - t_1) + \cdots + (t_i - t_{i-1} - 1) = t_i + 1$ derivations and some divisions by non-zero functions. By successive applications of Rolle's theorem we have that $\delta(u,\lambda)$ has less than t_i isolated roots on σ^i for all $\lambda \in W^i - \Sigma$.

This concludes the proof of point i) when u_0 is not a center.

In the center case, we use the formula (4.28) above: the roots of $\delta(u,\lambda)$, different from 0, are roots of

$$\sum_{j=1}^{\ell} a_{2j+1}(\lambda)h_{2j+1}(u,\lambda).$$

Next, taking a normal form for X_λ for a big order N, it can easily be seen that the factors $h_{2j+1}(u, \lambda) = k_j(U, \lambda)$ where $U = u^2$ and k_j is a C^k-function; taking N large enough, this order k can be taken arbitrarily large. Now $k_j(U, \lambda_0) = U^j(1 + O(U))$ and we can repeat the above proof to obtain the bound $s_\delta(e)$ for the cyclicity.

Point ii) If $\varphi_1, \ldots, \varphi_\ell$ are independent functions, assume that local coordinates $\lambda_1, \ldots, \lambda_\ell, \ldots, \lambda_\Lambda$ can be chosen in the parameter space, with $\varphi_i = \lambda_i$, $i = 1, \ldots, \ell$ and $\lambda_0 = (0, \ldots, 0)$.

Of course, we can always assume that the minimal system $\{\widetilde{\varphi}_1, \ldots, \widetilde{\varphi}_\ell\}$ is adapted to u_0 which is supposed to be translated to 0,

$$\delta(u, \lambda) = \sum_{i=1}^{\ell} \lambda_i \, h_i(u, \lambda),$$

with $h_i(u, 0) = u^{n_i}(1 + 0(u))$, $n_1 < n_2 \cdots < n_\ell$.
The essential point is that the sequence $\{h_1(u, 0), \ldots, h_\ell(u, 0)\}$ is a *Tchebychef system* (see [J1]). This means that there exists some $U > 0$ such that the function

$$\widetilde{\delta}(u, \widetilde{\lambda}) = \sum_{i=1}^{\ell} \lambda_i \, h_i(u, 0) \qquad \widetilde{\lambda} = (\lambda_1, \ldots, \lambda_\ell)$$

has at most $\ell - 1$ roots on $[0, U]$, counted with multiplicity. Moreover, the bifurcation diagram for the roots of $\widetilde{\delta}$ is the same as for the polynomial $\sum_{i=1}^{\ell} \lambda_i \, u^i$ on $[0, U]$.

In particular, we have a simplex $\Delta \subset S^{\ell-1}$ in the unit sphere of $\mathbb{R}^\ell$ such that for $\forall \widetilde{\lambda} \in \overset{\circ}{\Delta}$, $\widetilde{\delta}(u, \lambda)$ has $\ell - 1$ simple roots on $]0, U[$.

Now, let $\widetilde{\lambda} = u\overline{\lambda}$, $\overline{\lambda} \in S^{\ell-1}$ and $u \in \mathbb{R}^+$ and let $\lambda' = (\lambda_{\ell+1}, \ldots, \lambda_\Lambda)$.

$$\delta(u, \lambda) = u \Big[\sum_{i=1}^{\ell} \overline{\lambda}_i \, h_i(u, 0) + 0(\lambda) \Big]. \tag{4.35}$$

The equation $\delta(u, \lambda) = 0$ is equivalent to

$$\sum_{i=1}^{\ell} \overline{\lambda}_i \, h_i(u, \lambda) + 0(\lambda) = 0. \tag{4.36}$$

If we take an open, non-empty compact subset Q of Δ, such that $\overline{Q} \subset \overset{\circ}{\Delta}$, we deduce that the equation (4.36) has at least $\ell - 1$ simple roots in $]0, U[$ for $\lambda = u\overline{\lambda}$ $(\overline{\lambda}, u) \in S^{\ell-1} \times \mathbb{R}^+$ and u sufficiently small.

This proves that $\mathcal{C}ycl\,(X_\lambda, \gamma_{u_0}) \geq \ell - 1$ in this case (the proof also works in the case when u_0 is a center). $\qquad\qquad \square$

Remark 19 *It is easy to generalize the point ii). For instance, we can replace the condition "$\mathcal{I}$ regular" by the following one:*
"*The map $\lambda \to \varphi(\lambda)$ is locally surjective at $\lambda = \lambda_0$.*"

It would be interesting to have algebraic characterizations for ideals $\mathcal{I}$ with this property. As an example of such an ideal, we may consider $\mathcal{I} = \{\lambda_1^3, \lambda_2^5\}$.

4.3.4 Melnikov functions

In this section we will consider 1-parameter analytic families X_ε with $\varepsilon \in \mathbb{R}$ which unfold a vector field X_0 of center type. In this case, it is possible to expand $\delta(u, \varepsilon)$ in terms of ε,

$$\delta(u, \varepsilon) = \sum_{i=1}^\infty M_i(u)\varepsilon^i. \tag{4.37}$$

The functions $M_i(u)$ are analytic on σ. We call M_i the ith *Melnikov function.* The above series converges near each $u_0 \in \sigma$.

If the family is not identically trivial ($\delta(u, \varepsilon) \not\equiv 0$), then there exists k, such that $M_i(u) \equiv 0$ for $i \leq k - 1$ and $M_k(u) \not\equiv 0$. In this case, the Bautin Ideal of X_ε at $\varepsilon = 0$ is generated by $\varepsilon^k : \mathcal{I} = (\varepsilon^k)$.

Moreover, the equation $\{\delta(u, \varepsilon) = 0\}$ is equivalent to

$$M_k(u) + O(\varepsilon^2) = 0. \tag{4.38}$$

If $u_0 \in \sigma$, the cyclicity of γ_{u_0} is bounded by the order of M_k at u_0 (half the order minus one if u_0 is a center). This is of course a trivial particular case of Theorem 9.

We recall now how to compute the Melnikov functions. First, we can find an analytic function K defined in a neighborhood of a center or periodic orbit of a center-type vector field X_0 such that KX_0 is a Hamiltonian vector field (div $(KX_0) \equiv 0$). This means that there exists a first integral H for X_0 which verifies $-dH = +KX_0 \rfloor dx \wedge dy$. (It is trivial to find H, K along a closed orbit of X_0. At a center point, we can write X_0 in polar coordinates and look for $H(r, \theta)$, such that $H(-r, \theta + \pi) = H(r, \theta)$).

Call ω_ε the dual form of $X_\varepsilon : \omega_\varepsilon = X_\varepsilon \rfloor dx \wedge dy$. Let σ be a transversal interval, parametrized by the value $h = H(u)$. Then, it is well known that

$$M_1(h) = -\int_{\gamma_h} \omega_1 \quad , \text{ if } \quad \omega_1 = \left.\frac{\partial \omega}{\partial \varepsilon}\right|_{\varepsilon=0},$$

and γ_h is the periodic orbit of X_0 in $\{H = h\}$.

This formula was recently extended by J.P. Françoise [F] and S. Yakovenko [Y] at any Melnikov function M_i for linear perturbation

$$dH + \varepsilon\omega_1.$$

I give a more general version obtained by J.C. Poggiale [Po] for any 1-parameter family.

Let $\omega_\varepsilon = dH + \varepsilon \omega_1 + \cdots \varepsilon^{k+1} \omega_{k+1} + o(\varepsilon^{k+1})$ (for some k). We choose a domain U, invariant by the flow of X_0 and containing the transversal σ.

Proposition 6 *Under the above assumptions, we assume that $M_j(h) \equiv 0$ for $j \leq k$. Then*

$$M_{k+1}(h) = \int_{\gamma_h} \Big(\sum_{i=1}^{k} g_i \, \omega_{k+1-i} - \omega_{k+1} \Big),$$

where the analytic functions g_i, $i = 1, \ldots, k$ are defined inductively on U by

$$\omega_i - g_i \, dH = \sum_{j=1}^{i-1} g_j \, \omega_{k-j} + dR_i.$$

Proof. We need the following result:

Lemma 16 *Let ω be an analytic 1-form defined in the neighborhood U. Then $\int_{\gamma_h} \omega \equiv 0$ for $h \in \sigma$ if and only if there exist analytic functions g, R, such that in this neighborhood $\omega = g dH + dR$.*

Proof of Lemma 16. If γ_{h_0} is a closed orbit, one can adopt action-angle variables (H, θ) near γ_{h_0} (H near h_0 and $\theta \in S^1$). $\omega = g dH + b d\theta$, and $\int_{\gamma_H} \omega \equiv 0 \Longleftrightarrow$ $\int_0^{2\pi} b(H, \theta) d\theta \equiv 0$. Then there exists a function $R(H, \theta)$ such that $b(H, \theta) d\theta = dR$ and $\omega = g dH + dR$. If γ_{h_0} is the center, we use the same method: the function $R(H, \theta)$ is analytic in (x, y)-coordinate at the origin. $\square$

Return now to the proof of Proposition 6.

First, by integration of the formula $\omega_\varepsilon = dH + \varepsilon \omega_1 + 0(\varepsilon)$ we obtain that $M_1(h) = -\int_{\gamma_h} \omega_1$ (the usual Melnikov formula at order 1).

We now make the following induction hypothesis: there exist functions g_1, $g_2, \ldots, g_{k-1}$, analytic on U, such that for all $j = 1, \ldots, k$,

$$M_j(h) = \int_{\gamma_h} \Big(\sum_{i=1}^{j-1} g_i \, \omega_{j-i} - \omega_j \Big) \equiv 0.$$

Using this relation for $j = k$, and applying Lemma 16, we find two analytic functions on $U : g_k$, R_k such that

$$-\sum_{i=1}^{k-1} g_i \, \omega_{k-i} + \omega_k = g_k \, dH + dR_k,$$

and then, $\omega_k - g_k \, dH = \sum_{i=1}^{k-1} g_i \, \omega_{k-i} + dR_k.$

This proves that the functions g_k, R_k may be constructed by recurrence. Next a direct expansion gives

$$\left(1 - \sum_{i=1}^{k} g_i \, \varepsilon^i\right)\omega_\varepsilon = d\left(H - \sum_{i=1}^{k} \varepsilon^i \, R_i\right) + \varepsilon^{k+1}\left(\omega_{k+1} - \sum_{i=1}^{k} g_i \, \omega_{k+1-i}\right) + o(\varepsilon^{k+1}).$$

As H is a Morse function we can find an analytic diffeomorphism in U, $\mathrm{Id} + O(\varepsilon)$ which changes $H - \sum_{i=1}^{k} \varepsilon^i \, R_i$ in H and does not modify the coefficient in ε^{k+1}. The result for $M_{k+1}(h)$ follows by integration, as for $k = 1$. $\square$

Remark 20

1) In the linear case $dH + \varepsilon\omega_1$ the formula is simply $M_{k+1}(h) = \displaystyle\int_{\gamma_h} g_k \, \omega_1$ where the g_i are given inductively by:

$$g_i \, dH + dR_i = -g_{i-1} \, \omega_1.$$

2) Proposition 6 is clearly true for smooth families.

If now X_λ is any unfolding of a center-type vector field X_{λ_0} an important question is:

"Can we deduce the cyclicity for X_λ by computing Melnikov functions for 1-parameter subfamilies $X_{\lambda(\varepsilon)}$ where $\lambda(\varepsilon)$ is any analytic arc in the parameter space with $\lambda(0) = \lambda_0$?"

Of course, if γ is a periodic orbit or a center of X_{λ_0}

$$Cycl(X_\lambda, \gamma) \geq \mathrm{Sup}_{\lambda(\varepsilon)} \, Cycl(X_{\lambda(\varepsilon)}, \gamma).$$

The above question reduces to one of proving that we have the reverse inequality $Cycl(X_\lambda, \gamma) \leq \mathrm{Sup}_{\lambda(\varepsilon)} \, \mathcal{O}_{\lambda(\varepsilon)}$ where $\mathcal{O}_{\lambda(\varepsilon)}$ is the order of the first non-zero Melnikov function for $X_{\lambda(\varepsilon)}$. We can be more explicit. Suppose that $\{\widetilde{\varphi}_1, \ldots, \widetilde{\varphi}_\ell\}$ is a minimal set of generators for the Bautin Ideal. Then considering Theorem 9, it would suffice to prove that

There exists an analytic curve $\lambda(\varepsilon)$ through λ_0 such that order $(\varphi_\ell \circ \lambda)(0) <$ order $(\varphi_i \circ \lambda)(0)$, for any $i \neq \ell$.

In fact, if the above claim is true, $\mathcal{O}_{\lambda(\varepsilon)} \geq$ order $h_\ell(u_0) = s_\delta(u_0) \geq Cycl(X_\lambda, \gamma_{u_0})$ where $\gamma = \gamma_{u_0}$. This is trivially true when the Bautin Ideal is regular. It is not known if this is true in general.

4.3.5 Application to quadratic vector fields

4.3.5.1 Bautin result

If we are interested in quadratic vector fields with at least one limit cycle, it is sufficient to consider the Kaypten-Dulac family X_λ, with a focus or center point at the origin. Using a rotation, we can eliminate one parameter and study the following 6-parameter family,

$$X_\lambda \begin{cases} \dot{x} &=& -y + \lambda_1 x - \lambda_3\, x^2 + (2\lambda_2 + \lambda_5)xy + \lambda_6\, y^2 \\ \dot{y} &=& x + \lambda_1 y + \lambda_2\, x^2 + (2\lambda_3 + \lambda_4)xy - \lambda_2\, y^2. \end{cases} \tag{4.39}$$

For any $\lambda \in \mathbb{R}^6$, the origin is a focus or a center. We can consider the return map on the $0x$-axis, and the difference function $\delta(x, \lambda) = \sum_{i=1}^{\infty} a_i(\lambda)x^i$.

It is easy to compute $a_1(\lambda) = e^{2\pi\lambda_1} - 1$.

The other seven first coefficients were computed in, for instance, [Du1] (see also [B], [Ye],...). We already know that it suffices to compute coefficients of odd order:

$$\begin{aligned}
a_3(\lambda) &=& c_3\,\lambda_5(\lambda_3 - \lambda_6) \mod (a_1) \\
a_5(\lambda) &=& c_5\,\lambda_2\,\lambda_4(\lambda_3 - \lambda_6)(\lambda_4 + 5(\lambda_3 - \lambda_6)) \mod (a_1, a_3) \\
a_7(\lambda) &=& c_7\,\lambda_2\,\lambda_4(\lambda_3 - \lambda_6)^2\,(\lambda_3\,\lambda - 2\lambda_6^2 - \lambda_2^2) \mod (a_1, a_3, a_5),
\end{aligned}$$

for some constant c_3, c_5, $c_7 \neq 0$.

A difficult result of Bautin in [B] is that the ideal generated by the coefficient $\mathcal{I}(a_i)$ in the ring of analytic functions of $\mathbb{R}^6$ is generated by a_1, a_3, a_5, a_7 (see [Y] for a geometrical proof). Hence, the Bautin Ideal at each λ_0 is generated by the germs of

$$v_1 = \lambda_1,\ v_3 = \lambda_5(\lambda_3 - \lambda_6),\ v_5 = \lambda_2\,\lambda_4(\lambda_3 - \lambda_6)(\lambda_4 + 5(\lambda_3 - \lambda_6))$$
$$\text{and}\ \ v_7 = \lambda_2\,\lambda_4(\lambda_3 - \lambda_6)^2\,(\lambda_3\,\lambda_6 - 2\lambda_6^2 - \lambda_2^2).$$

Using Proposition 4 and Lemma 14, we can write locally near $x = 0$ and each $\lambda_0 \in \mathbb{R}^6$,

$$\delta(x, \lambda) = v_1\,h_1(x, \lambda) + v_3\,h_3(x, \lambda) + v_5\,h_5(x, \lambda) + v_7\,h_7(x, \lambda)$$

$$\text{with}\ \ h_i = x^i(1 + \psi_i(x, \lambda))\ \text{and}\ \varphi_i = 0(x).$$

This formula implies that $s_\delta(0) \leq 3$ at each $\lambda_0 \in \mathbb{R}^6$, so that, by Theorem 9, at most three limit cycles can bifurcate from the origin by perturbations of the parameter λ.

The set $Z = \{v_1 = v_3 = v_5 = v_7 = 0\}$ is the set of parameter values for which X_λ is of center-type. In this case the origin is a center surrounded by a "center basin" B_λ of periodic orbits. This set was first described by Dulac [Du1].

It is an algebraic subset of $\mathbb{R}^6$ with four irreducible components:

$$
\begin{aligned}
Q^H &= \{\lambda_1 = \lambda_4 = \lambda_5 = 0\} \\
Q^R &= \{\lambda_1 = \lambda_2 = \lambda_5 = 0\} \\
Q^D &= \{\lambda_1 = \lambda_3 - \lambda_6 = 0\} \\
Q^M &= \{\lambda_1 = \lambda_5 = \lambda_4 + 5(\lambda_3 - \lambda_6) = \lambda_3\,\lambda_6 - 2\lambda_6^2 - \lambda_2^2 = 0\}.
\end{aligned}
$$

Each of these components has a geometrical meaning; Q^H consists of the Hamiltonian vector fields (equation: div $X_\lambda \equiv 0$), Q^R consists of the reversible ones: each has a symmetrical phase portrait with respect to a line through the origin, vector fields in Q^D contain three invariant lines (real or complex), and vector fields in Q^M contain an invariant parabola and an invariant cubic. Moreover, for each center one has an explicit first integral and an explicit integrating factor (see [Du1] and [S1] for a more recent and complete study).

Outside the intersection of components, the set Z is a submanifold and the ideal of Bautin is regular.

Consider for instance $\lambda_0 \in Q^M \backslash Q^H \cup Q^R \cup Q^D$.

At such a point $\lambda_4 \neq 0$, $\lambda_2 \neq 0$, $\lambda_3 - \lambda_6 \neq 0$.

Then, the ideal $\mathcal{I}^{\lambda_0}$ is generated by λ_1, λ_5, $\lambda_4 + 5(\lambda_3 - \lambda_6)$ and $\lambda_3\,\lambda_6 - 2\lambda_6^2 - \lambda_2^2$, so that $\mathcal{I}^{\lambda_0}$ is regular and $\ell(\mathcal{I}^{\lambda_0}) = 4$. Applying point ii) of Theorem 2, we see that the cyclicity at such a value λ_0 is not smaller than three and hence equal to three: one can find values of λ near λ_0 with three "small" limit cycles.

In the same way one can prove that the cyclicity is greater than 2, greater than 2, and greater than 1, at the regular points of Q^H, Q^R and Q^D respectively.

We have seen that Bautin obtained an upper bound for the index s_δ at the origin for any $\lambda_0 \in Z$. For orbits of X_{λ_0} different from the center point no general result is known, except for some particular cases: for instance Horozov and Iliev [HI] proved that $s_\delta(u_0) = 2$ for the open subset of Q^H corresponding to generic Hamiltonian vector fields with three saddle points and one center, for any $u_0 \in \sigma$, transversal interval in the center basin. As we know that the cyclicity is not less than two, this implies that $\mathcal{C}ycl(X_\lambda, \gamma_{u_0}) = 2$ in this case. For such a Hamiltonian vector field, the basin B_λ is a disk bounded by a saddle connection and the authors obtained the whole bifurcation diagram. In particular, they proved that the total number of limit cycles for near by vector fields is at most two. This includes the study of the saddle connection bifurcation we want to consider in the next chapter.

4.3.5.2 Bogdanov-Takens unfolding

Some unfoldings can be reduced to perturbations of center-type vector fields. For instance, let us consider again the Bogdanov-Takens unfolding introduced in Chapter 1, written in the normal form:

$$
X^+_{\mu,\nu,\lambda} \begin{cases} \dot{x} &= y \\ \dot{y} &= x^2 + \mu + y(\nu + x) + yx^2\ h(x,\lambda) + y^2\ Q(x,y,\lambda), \end{cases}
$$

where h, Q are smooth functions. To reduce this unfolding to a perturbation of a center-type vector field, we can use the following rescaling formulas:

$$x = \varepsilon^2 \, \bar{x}, \; y = \varepsilon^3 \, \bar{y}, \; \mu = -\varepsilon^4 \quad \nu = \varepsilon^2 \bar{\nu}. \tag{4.40}$$

Taking $(\bar{x}, \bar{y})$ belonging to some compact domain $\overline{D}$ in $\mathbb{R}^2$ and $\bar{\nu}$ belonging to a close interval K to be defined below, $\varepsilon \in \mathbb{R}^+$, these formulas transform the family X^+ into a new one $\overline{X}_{\bar{\nu}, \varepsilon, \lambda} = \dfrac{1}{\varepsilon} \, \widehat{X}$, where $\widehat{X}$ is the family X^+ written in coordinates $(\bar{x}, \bar{y})$,

$$\overline{X}_{\bar{\nu}, \varepsilon} \begin{cases} \dot{\bar{x}} &= \bar{y} \\ \dot{\bar{y}} &= \bar{x}^2 - 1 + \varepsilon \bar{y}(\bar{\nu} + \bar{x}) + O(\varepsilon^2). \end{cases} \tag{4.41}$$

This is an ε-perturbation of the Hamiltonian vector field X_0 with the Hamiltonian function $H_1 = \dfrac{1}{2} \, \bar{y}^2 + \bar{x} - \dfrac{\bar{x}^3}{3}$ (see Figure 4.2:).

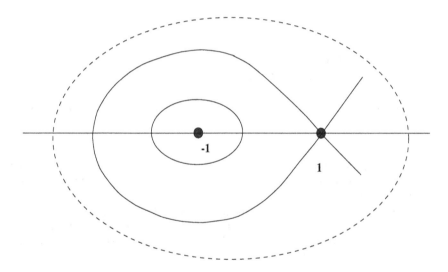

Figure 4.2

To study limit cycles bifurcating from the Hamiltonian cycles we take a disk $\overline{D}$ big enough to contain the disk bounded by the homoclinic loop through the saddle $s = (+1, 0)$. For each $h \in \left[-\dfrac{2}{3}, \dfrac{2}{3}\right]$, let γ_h be the cycle in $\{H = h\}$; γ_0 is the center $(-1, 0)$ and $\gamma_{2/3}$ the homoclinic loop. Let $\omega_{\nu, \varepsilon}$ be the dual form of $X_{\nu, \varepsilon}$,

$$\omega_{\nu, \varepsilon} = dH - \varepsilon(\nu + x)ydx + O(\varepsilon^2). \tag{4.42}$$

From Proposition 5, the displacement function for the family $\overline{X}_{\nu,\varepsilon}$ has the following expansion in ε:

$$\delta(h, \nu, \varepsilon) = \varepsilon M_1(h, \nu) + o(\varepsilon),$$

where

$$M_1(h, \nu) = \int_{\gamma_h} (\nu + x) y dx = \nu I_0(h) + I_1(h). \tag{4.43}$$

We write

$$I_i(h) = \int_{\gamma_h} x^i \, y dx. \tag{4.44}$$

$I_0(h)$ is equal to the area of the disk bounded by γ_h.

Hence $I_0(h) > 0$ for $h \neq 0$ and $I_0(h) \sim h + \frac{2}{3}$.

As $I_1(0) = 0$, the function $B_1(h) = \dfrac{I_1}{I_0}(h)$ is defined and analytic for $h \in [-\frac{2}{3}, \frac{2}{3}[$. The equation for limit cycles $\{\delta = 0\}$ is equivalent to

$$\frac{\delta}{\varepsilon I_0} = \nu + B_1(h) + O(\varepsilon) = 0, \tag{4.45}$$

and the unicity of limit cycle in the Bogdanov Takens unfolding as claimed in Chapter 1, reduces to the proof of the following result:

Theorem 10 *(Bogdanov): For all $h \in \left[-\dfrac{2}{3}, \dfrac{2}{3}\right[$, $B_1'(h) > 0$ and $B_1'(h) \to \infty$ as $h \to \dfrac{2}{3}$.*

We will prove this result in the next section about abelian integrals. The fact that $B_1'(0) > 0$ implies that the line H which we introduced in Chapter 1 is a line of generic Hopf bifurcations.

We will study the line of homoclinic loops in the next chapter. Here, the theorem, applied for $h \in [-\frac{2}{3}, h_1]$ where $h_1 \in [-\frac{2}{3}, \frac{2}{3}]$ is chosen near $\frac{2}{3}$, implies the unicity of the limit cycle in the interior of the tongue T between the two lines H, C.

4.3.5.3 An example of a non-regular ideal

In the above example, the computation of cyclicity reduces to the computation of a Melnikov function. This is because the Bautin Ideal was regular (generated by ε in the Bogdanov-Takens unfolding). In this section, we will consider a case of a non-regular ideal for the quadratic family in the Kaypten-Dulac form (4.39). This means that we will choose λ_0 at the intersection of center components. Let $\lambda_0 \in Q^H \cap Q^R - \{0\}$. This means that $\lambda_0 = (0, \ldots, 0, \lambda_6)$ with $\lambda_6 \neq 0$. Changing $(x, y) \to (\beta y, \beta x)$ with $\beta = | \lambda_6 |^{-1/2}$ and $t \to sign(\lambda_6)t$, we can assume that $\lambda_6 = -1$. The vector field X_{λ_0} is a Hamiltonian, with Hamiltonian function

$H_2(x, y) = \frac{1}{2} y^2 + \frac{1}{2} x^2 + \frac{1}{3} x^3$ similar to the one in Section 3.5.2, with the interval $[-\frac{2}{3}, \frac{2}{3}]$ replaced by the interval $[0, \frac{1}{6}]$.

At the parameter value $\lambda_0 = (0, \ldots, 0, -1)$, the Bautin Ideal is generated by the germs of $\lambda_1, \lambda_5, \lambda_2\lambda_4$. Hence, as a consequence of Proposition 4, we can divide $\delta(h, \lambda)$ in the ideal, near (h_0, λ_0), for any $h_0 \in [0, 1/6[$:

$$\delta(h, \lambda) = \lambda_1 h_1(h, \lambda) + \lambda_5 h_2(h, \lambda) + \lambda_2\lambda_4 h_3(h, \lambda). \tag{4.46}$$

Clearly, $\{\lambda_1, \lambda_5, \lambda_2\lambda_4\}$ is a minimal set of generators at λ_0. The functions $H_i(h) = h_i(h, \lambda_0)$ are analytic on $[0, 1/6[$. To compute them, we can use some 1-parameter subfamilies of (4.39), in which we have put $\lambda_6 = -1$ and made the other changes indicated above:

– Computation of H_1.

Consider the subfamily

$$\lambda(\varepsilon) = (\lambda_1 = \varepsilon, \lambda_2 = \cdots = \lambda_5 = 0, \lambda_6 = -1),$$

$$X_{\lambda(\varepsilon)} \begin{cases} \dot{x} &= y + \varepsilon y \\ \dot{y} &= -x - x^2 + \varepsilon y. \end{cases} \tag{4.47}$$

Substituting $\lambda(\varepsilon)$ into (4.39) gives

$$\delta(h, \lambda(\varepsilon)) = \varepsilon H_1(h) + O(\varepsilon).$$

This shows that $H_1(h) = M_1(h)$, the first Melnikov function for the family (4.47). Its dual 1-form is

$$\omega_\varepsilon = dH + \varepsilon(-y dx + x dy).$$

From Proposition 5, we obtain

$$M_1(h) = -\int_{\gamma_h} - y dx + x dy = 2 I_0(h) \tag{4.48}$$

(with notation $I_i(h)$ introduced in (4.44)).

– Computation of H_2.

Take

$$\lambda(\varepsilon) = (\lambda_5 = \varepsilon, \lambda_1 = \lambda_2 = \lambda_3 = \lambda_4 = 0, \lambda_6 = -1),$$

$$X_{\lambda(\varepsilon)} = \begin{cases} \dot{x} &= y \\ \dot{y} &= -x - x^2 + \varepsilon xy. \end{cases}$$

Then

$$H_2(h) = M_1(h) = -I_1(h). \tag{4.49}$$

– Computation of H_3.

Consider the family

$$\lambda(\varepsilon) = (\lambda_1 = \lambda_3 = \lambda_5 = 0, \lambda_2 = \lambda_4 = \varepsilon, \ \lambda_6 = -1),$$

$$X_{\lambda(\varepsilon)} \begin{cases} \dot{x} &= y + \varepsilon(y^2 + xy - x^2) \\ \dot{y} &= -x - x^2 + 2\varepsilon xy. \end{cases} \tag{4.50}$$

Now, $\delta(h, \lambda(\varepsilon)) = \varepsilon^2 \, H_3(h) + 0(\varepsilon^2)$ and $H_3(h) = M_2(h)$, the second Melnikov function for (4.50). Its dual 1-form is

$$\omega_\varepsilon = dH + \varepsilon\bar{\omega} \quad \text{with} \quad \bar{\omega} = -2xy dx + (y^2 + xy - x^2)dy. \tag{4.51}$$

We can verify that $M_1(h) \equiv 0$:

$$M_1(h) = \int_{\gamma_h} -2xy dx + (y^2 + xy - x^2)dy = -\int_{\gamma_h} d\left(yx^2 + \frac{y^3}{3}\right) + \int_{\gamma_h} yx dy,$$

and

$$\int_{\gamma_h} xy dy = \int_{\gamma_h} x\left(dH - (x + x^2)dx\right) \equiv 0.$$

As was proved in 3.4, there exist functions g, R such that

$$\bar{\omega} = g dH + dR. \tag{4.52}$$

This equation for g is equivalent to

$$d\bar{\omega} = dg \wedge dH \quad , \text{ i.e.,}$$

$$-y dx \wedge dy = dg \wedge [y dy + (x + x^2)dx]. \tag{4.53}$$

Clearly $g(x, y) = x$ is a solution for (4.53), and by Proposition 6, we have

$$H_3(h) = M_2(h) = \int_{\gamma_h} x\bar{\omega} = \int_{\gamma_h} -2yx^2 dx + x(y^2 + xy - x^2)dy. \tag{4.54}$$

We want to compute the four integrals in (4.54). To simplify notation, we write $\alpha \sim 0$ for $\int_{\gamma_h} \alpha \equiv 0$.

$$
\begin{aligned}
&1) \quad yx^2 \, dx \sim y(dH - y dy - x dx) \sim -yx dx = -\omega_1 &&(4.55)\\
&2) \quad\quad yx^2 dy \sim x^2(dH - (x + x^2)dx) \sim 0 &&(4.56)\\
&3) \quad x^3 dy \sim (3H - \frac{3}{2}y^2 - \frac{3}{2}x^2)dy \sim -\frac{3}{2}x^2 \, dy\\
&\quad\quad \text{and } x^2 \, dy \sim -y d(x^2) = -2xy dx = -2\omega_1,
\end{aligned}
$$

so that
$$x^3 dy \sim 3\omega_1. \tag{4.57}$$

4) $xy^2 \, dy \sim 2x(H - \frac{1}{2} x^2 - \frac{1}{3} x^3)dy \sim -2h\omega_0 - 3\omega_1 - \frac{2}{3} x^4 \, dy$

$$x^4 \, dy \sim -4yx^3 \, dx \sim -4y(3H - \frac{3}{2} y^2 - \frac{3}{2} x^2)dx$$

$$\sim -4h\omega_0 + 6y^3 \, dx + 6yx^2 \, dx \sim -4h\omega_0 - 6\omega_1 + 6y^3 \, dx$$

$$\text{and } y^3 \, dx \sim -3y^2 \, xdy.$$

Then, finally,

$$xy^2 \, dy \sim -2h\omega_0 - 3\omega_1 - \frac{2}{3} \left(-4h\omega_0 - 6\omega_1 - 18xy^2 \, dy \right)$$

$$-11xy^2 \, dy \sim \frac{2}{3} h\omega_0 + \omega_1. \tag{4.58}$$

Collecting the different contributions (4.55)-(4.58), we obtain:

$$y\bar{\omega} \sim -\frac{2}{33} h\omega_0 + \frac{20}{11} \omega_1 \tag{4.59}$$

and the following expression for $M_3(h)$:

$$H_3(h) = M_2(h) = -\frac{2}{33} hI_0(h) + \frac{20}{11} I_1(h). \tag{4.60}$$

Taking the generators $\varphi_1 = -2\lambda_1$, $\varphi_2 = -\lambda_5$, $\varphi_3 = -\frac{2}{33} \lambda_2 \lambda_4$ for the Bautin Ideal at λ_0, we have

$$\delta(h, \lambda) = \varphi_1 \, h_1(h, \lambda) + \varphi_2 \, h_2(h, \lambda) + \varphi_3 \, h_3(h, \lambda) \tag{4.61}$$

for new factors h_i, such that $H_1(h) = I_0$, $H_2(h) = I_1$, and $H_3(h) = hI_0 - 3I_1$.

Proposition 7 *For each*

$$h \in \,]0, 1/6[, \quad \det \begin{pmatrix} H_1 & H_1' & H_1'' \\ H_2 & H_2' & H_2'' \\ H_3 & H_3' & H_3'' \end{pmatrix} \neq 0. \tag{4.62}$$

Proof. In order to verify (4.62) we can replace the functions H_1, H_2, H_3 by the multiples LH_1, LH_2, LH_3, where $L(h)$ is an analytic function on $]0, \frac{1}{6}[$, everywhere non-zero. As $I_0(h) > 0$, for $h \in]0, 1/6[$ we can take $L = I_0^{-1} = H_1^{-1}$. Let $B_2(h) = \frac{I_1}{I_0}(h)$ as above. Then (4.62) is equivalent to

$$\begin{vmatrix} 1 & 0 & 0 \\ B_2 & B_2' & B_2'' \\ B_2 + h & B_2' + 1 & B_2'' \end{vmatrix} \neq 0, \; i.e., \; B_2''(h) \neq 0, \; for \; \forall h \in [0, 1/6[.$$

We will prove this result in the next section.   □

A consequence of Proposition 7 is that $s_\delta(h) = 2$ for $\forall h \in]0, 1/6[$. At the center point $(0,0)$, corresponding to $h = 0$, we have $h \sim x^2$. Then the fact that $B_2''(0) \neq 0$ (see next section) will clearly imply that $s_\delta(0) = 2$.

As a consequence of Theorem 9, it now follows that

$$\mathcal{C}ycl(X_\lambda, \gamma_{h_0}) \geq 2 \quad \text{for } \forall h \in [0, 1/6[.$$

In fact, $\mathcal{C}ycl(X_\lambda, \gamma_h) = 2$. This can be proved using the remark after the proof of Theorem 2, or deduced from the above result [HI]. According to this paper we can find a sequence $(\lambda_i)_i \to \lambda_0$, $\lambda_i \in Q^H \backslash Q^R$ such that for each i, X_{λ_i} has a periodic orbit γ_i with cyclicity 2 and $\gamma_i \to \gamma_{h_0}$. Then, using the semi-continuity of the cyclicity proved in Lemma 2.3, we obtain that $\mathcal{C}ycl(X_\lambda, \gamma_{n_0}) \geq 2$ and so that the cyclicity is 2. We have proved

Theorem 11 Let $\lambda_0 \in Q^R \cap Q^R - \{0\}$ and γ_{h_0}, $h_0 \in [0, 1/6[$ be any regular limit periodic set for X_{λ_0}. Then in the Kaypten-Dulac family X_λ, $\mathcal{C}ycl(X_\lambda, \gamma_{h_0}) = 2$.

Remark 21 From formula (4.46) and Proposition 7, it is possible to deduce the bifurcation diagram for X_λ, near λ_0.

4.3.6 Some properties of Abelian Integrals

In the preceding sections we have seen that the properties of a small deformation of a center-type vector field are closely related to the properties of Abelian integrals. Abelian integrals are integrals of algebraic 1-forms of the cycles of a Hamiltonian function $H: \int_{\gamma_n} \bar{\omega}$. In our applications H and $\bar{\omega}$ are rather special. In particular H may be reduced to the form $H(x, y) = y^2 - P(x)$ where $P(x) = \sum_{i=0}^{2g+1} p_i \, x^i$.

To begin with, we want to present briefly this special case. A general result was proved by Petrov [Pe].

We assume that a continuous family of closed curves γ_h is chosen, each of them in the level $\{H = h\}$. To make the theory general one has to choose h in the universal covering of $\mathbb{C} - \Sigma$ where Σ is the set of critical values of H, but in our applications, it will suffice to take h in some interval (the image by H of an interval between a center e and a next critical point s). This interval is contained in $\mathbb{R}$.

For each meromorphic 1-form ω we can consider the Abelian integral $I_\omega(h) = \int_{\gamma_\lambda} \omega$. In particular, let

$$\omega_i = x^i \, y dx \;, \quad \alpha_i = \frac{x^i}{y} \, dx, \quad \text{and} \quad I_i(h) = \int_{\gamma_h} \omega_i \;, \quad J_i = \int_{\gamma_h} \alpha_i.$$

Now, the two most important results about Abelian integrals for $H(x, y) = y^2 - P(x)$ are

1) For each algebraic ω, there exist $2g$ polynomials in h: $Q_i(h)$, $i = 0, \ldots, 2g-1$, such that

$$I_\omega(h) = \sum_{i=0}^{2g-1} Q_i(h) I_i(h).$$

2) There exist two $(2g - 1) \times (2g - 1)$ matrices C and M such that

$$I = hJ + CJ - MI \quad \text{and} \quad \frac{d}{dh} I = -\frac{1}{2} J, \tag{4.63}$$

where $I = (I_i)_{i=0,\ldots,2g-1}$, $J = (J_i)_{i=0,\ldots,2g-1}$.
One can eliminate J to obtain the linear differential system

$$(\text{Id} + M)I = -2 (h \ \text{Id} + C) \frac{dI}{dt}. \tag{4.64}$$

This system, which is singular at critical values of H, is precisely the Gauss-Manin connexion of H.

We are going to give a short proof of point 2). This proof, communicated to me by S. Yakovenko, is based on a preprint of Givental.

First, we have $d\omega_i = d(x^i \, ydx) = x^i \, dy \wedge dx$ so that $d\omega_i = -\frac{1}{2} \frac{x^i}{y} \, dx \wedge dH = -\frac{1}{2} \alpha_i \wedge dH$, $\alpha_i = 0, \ldots, 2g - 1$.

This is equivalent to $\dfrac{dI}{dh} = -\dfrac{1}{2} J$.
Next, for any $n = 0, \ldots, 2g - 1$,

$$\omega_n = x^n \, ydx = \frac{x^n(P+h)}{y} \, dx = h\alpha_n + \frac{x^n \, P}{y} \, dx. \tag{4.65}$$

Dividing $x^n \, P$ by P' we have

$$x^n \, P = \sum_{i=0}^{2g-1} C_{ni} \, x^i + Q(x) P'(x). \tag{4.66}$$

This formula defines the polynomial $Q(x)$, of degree $\deg Q = n+2g+1-2g = n + 1$. Let

$$Q(x) = \sum_{j=0}^{n+1} q_{nj} \, x^j. \tag{4.67}$$

Substituting (4.66) in (4.65), we get

$$\omega_n = x^n \, ydx = h\alpha_n + \sum_{i=0}^{2g-1} C_{ni} \, \alpha_i + \frac{QP'}{y} \, dx. \tag{4.68}$$

Now, writing $\Omega \sim 0$ for $\displaystyle\int_{\gamma_n} \Omega \equiv 0$, and using $P'dx \sim 2ydy$ in (4.68), we have

$$\omega_n \sim h\alpha_n + \sum_{i=0}^{2g-1} C_{ni} \, \alpha_i + 2Qdy. \tag{4.69}$$

But

$$2Qdy \sim -2Q'ydx = -2\sum_{i=0}^{n} (i+1)q_{ni+1} \, \omega_i, \tag{4.70}$$

so that

$$\omega_n = h\alpha_n + \sum_{i=0}^{2g-1} C_{ni} \, \alpha_i - 2\sum_{i=0}^{n} (i+1)q_{ni+1} \, \omega_i. \tag{4.71}$$

Putting

$$C = (C_{ni})_{n,i} \quad \text{and} \quad M = 2((i+1)q_{ni+1})_{n,i}, \tag{4.72}$$

(4.71) is the system (4.63).

We want to apply these general considerations to obtain the results about the ratio $\frac{I_1}{I_0}$ claimed in Theorem 10 and in the proof of Proposition 7 above.

First, notice that the Hamiltonian $H_1(x_1, y_1) = \dfrac{1}{2} \, y_1^2 + x_1^2 - \dfrac{x_1^3}{3}$ in Section 3.5.2 and the Hamiltonian $H_2(x_2, y_2) = \dfrac{1}{2} \, y_2^2 + \dfrac{1}{2} \, x_2^2 + \dfrac{x_2^3}{3}$ are equivalent to the Hamiltonian $H(x, y) = y^2 - x + x^3$. If we put

$$\begin{cases} x_2(x) & = & \frac{\sqrt{3}}{2}x - \frac{1}{2} \\ y_2(x) & = & \sqrt{2}y, \end{cases} \tag{4.73}$$

then

$$H_2(x_2, y_2) = \frac{\sqrt{3}}{8} \, H(x, y) + \frac{1}{12}. \tag{4.74}$$

If

$$\begin{cases} x_1(x) & = & -\sqrt{3}x \\ y_1(x) & = & \sqrt{2}y, \end{cases} \tag{4.75}$$

then we have

$$H_1(x_1, y_1) = \sqrt{3} \, H(x, y), \tag{4.76}$$

so that the ratios $B_1(h_2)$, $B_2(h)$ of $\dfrac{I_1}{I_0}$ for the Hamiltonian functions H_1, H_2 are simply related to the ratio B for H,

$$B_1(\sqrt{3}h) = -\sqrt{3}B(h) \quad \text{and} \quad B_2\left(\frac{\sqrt{3}}{8} h + \frac{1}{12}\right) = \frac{3}{4}\sqrt{2}B(h) - \frac{1}{2}. \tag{4.77}$$

In order to prove the claims in Theorem 10 and of Proposition 7, we only need to prove the following theorem for the analytic function:

$$B(h) = \frac{I_1}{I_0}(h) : \left[-\frac{2}{3\sqrt{3}}, \frac{2}{3\sqrt{3}} \right[\to \mathbb{R}. \tag{4.78}$$

Theorem 12 $B'(h) < 0$, *for* $\forall h \in \left[-\frac{2}{3\sqrt{3}}, \frac{2}{3\sqrt{3}} \right[$, *and* $B'(h) \to -\infty$, *for* $h \to \frac{2}{3\sqrt{3}}$. *Moreover* $B''(h) < 0$, *for* $\forall h \in \left[-\frac{2}{3\sqrt{3}}, \frac{2}{3\sqrt{3}} \right[$.

Proof. First, we want to prove that the function $B(h)$ verifies the Ricatti equation

$$9\left(\frac{4}{27} - h^2 \right) \frac{dB}{dh} = -7B^2 - 3hB + \frac{5}{3}. \tag{4.79}$$

This is an easy consequence of the general formula (4.63). Repeating the above proof we have here $P(x) = x - x^3$, and then

$$\omega_0 = y dx = \frac{h + P}{y} dx = h\alpha_0 + \frac{P dx}{y}. \tag{4.80}$$

Writing $P(x) = \frac{2}{3} x + \frac{1}{3} x(1 - 3x^2)$ and substituting in (4.80), we obtain

$$\frac{5}{3} \omega_0 \sim h\alpha_0 + \frac{2}{3} \alpha_1. \tag{4.81}$$

Similarly,

$$xP(x) = \frac{2}{9} + (-\frac{2}{9} + \frac{2}{3} x^2)(1 - 3x^2) \quad \text{and}$$

$$\omega_1 \sim h\alpha_1 + \frac{2}{9} \alpha_0 - \frac{4}{3} \omega_1, \tag{4.82}$$

so

$$\frac{7}{3} \omega_1 \sim \frac{2}{9} \alpha_0 + h\alpha_1. \tag{4.83}$$

Relations (4.81), (4.83) with $\frac{dI_i}{dh} = -\frac{1}{2} J_i$ give

$$\begin{cases} \frac{5}{3} I_0 = 2hI_0' + \frac{4}{3} I_1' \\ \frac{7}{3} I_1 = \frac{4}{9} I_0' + 2hI_1', \end{cases} \tag{4.84}$$

which can be solved in I_0', I_1',

$$\begin{cases} (\frac{4}{27} - h^2)I_0' = -\frac{5}{6} hI_0 + \frac{7}{9} I_1 \\ (\frac{4}{27} - h^2)I_1' = \frac{5}{27} I_0 - \frac{7}{6} hI_1. \end{cases} \tag{4.85}$$

Notice that the roots of $\dfrac{4}{27} - h^2$: $\pm\dfrac{2}{3\sqrt{3}}$ are the critical values of $H(y = 0)$. Now
(4.85) implies the Ricatti equation (4.79) for $B(h) = \dfrac{I_1}{I_0}(h)$.

This equation for $B(h)$ means that the graph of this function belongs to an orbit of the following vector field Z on the space $\mathbb{R}^2$ of coordinates (h, B):

$$Z = 9\left(\frac{4}{27} - h^2\right) \frac{\partial}{\partial h} + \left(-7B^2 - 3hB + \frac{5}{3}\right) \frac{\partial}{\partial B}. \qquad (4.86)$$

This vector field has four critical points:

$$\begin{aligned}
\alpha_0 &= \left(-\frac{2}{3\sqrt{3}}, \frac{1}{\sqrt{3}}\right), \ \alpha_1 = \left(\frac{2}{3\sqrt{3}}, \frac{5}{7\sqrt{3}}\right), \\
\alpha_0' &= \left(-\frac{2}{3\sqrt{3}}, -\frac{5}{7\sqrt{3}}\right), \ \alpha_1' = \left(\frac{2}{3\sqrt{3}}, -\frac{1}{\sqrt{3}}\right),
\end{aligned}$$

and admits the lines $\Delta_0 = \left\{h = -\dfrac{2}{3\sqrt{3}}\right\}$ and $\Delta_1 = \left\{h + \dfrac{2}{3\sqrt{3}}\right\}$ as invariant lines. Along these lines Z is normally hyperbolic and in restriction to Δ_0 and Δ_1 the critical points are also hyperbolic.

The four critical points are hence hyperbolic and it is easily checked that α_0 and α_1' are saddle points, while α_0' and α_1 are unstable and stable nodes, respectively. The phase portrait of Z in the vertical strip $U = \left\{B \geq 0, \ -\dfrac{2}{3\sqrt{3}} \leq h \leq \dfrac{2}{3\sqrt{3}}\right\}$ can now easily be obtained taking into account the value of the vertical component of Z when $B = 0$ and when B is large (see Figure 4.3).

In particular, we notice the existence of a unique Z-orbit lying in the interior of U and having the saddle point $\alpha_0 = (-\frac{2}{3\sqrt{3}}, -\frac{1}{\sqrt{3}})$ as an α-limit point: it is the unstable separatrix Γ of α_0, which tends to α_1, for $t \to +\infty$.

Noticing that $B(h) \to \dfrac{1}{\sqrt{3}}$, for $h \to -\dfrac{2}{3\sqrt{3}}$, it follows that *the graph of $B(h)$ is equal to Γ* (of course, this implies that $B(h) \to \dfrac{5}{7\sqrt{3}}$ for $h \to \dfrac{2}{3\sqrt{3}}$).

Let us show that $B'(h) < 0$, for all $h \in \left[-\dfrac{2}{3\sqrt{3}}, \dfrac{2}{3\sqrt{3}}\right[$.

For $h = -\dfrac{2}{3\sqrt{3}}$, we have that $B'(-\dfrac{2}{3\sqrt{3}}) = -\dfrac{1}{8}$. This is simply obtained, computing the slope of the eigenspace at the saddle point α_0. For the other values of h, we make the following qualitative reasoning. We consider the equation $\{-7B^2 - 3hB + \dfrac{5}{3} = 0\}$, giving the points where Z is horizontal. This equation defines a hyperbola, whose two connected components are graphs of functions of h. The part of this hyperbola contained in the strip U is an arc S joining α_0 and α_1.

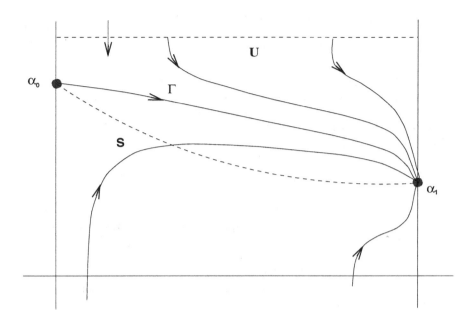

Figure 4.3

Along S, we can solve h in terms of B: $h = \dfrac{-7B^2 + 5/3}{3B}$ (since $B \neq 0$ on S). Hence Z is transverse to S and directed to the right. We now study the position of S with respect to Γ. At α_0, the tangent to S has a slope equal to $-\dfrac{1}{4}$, which is smaller than the slope $B'\left(-\dfrac{2}{3\sqrt{3}}\right) = -\dfrac{1}{8}$ of Γ at the same point. Then, in the neighborhood of α_0, the separatrix Γ is above S. But as, along S, the vector field Z is transverse to S and directed to the right, the orbit Γ is not allowed to cut S again for $t \to +\infty$. The orbit Γ is then entirely located in U, above S. But in this region, the vertical component of Z is negative. It follows that $B'(h) < 0$, for all $h \in \left[-\dfrac{2}{3\sqrt{3}} , \dfrac{2}{3\sqrt{3}}\right[$. For $h \to \dfrac{2}{3\sqrt{3}}$, $B'(h) \to -\infty$ because an easy computation gives that the eigenvalue at α_1 along Δ_1 is greater than the transversal eigenvalue (the two eigenvalues are negative).

Let us now show that $B''(h) < 0$, for all $h \in \left[-\dfrac{2}{3\sqrt{3}} , \dfrac{2}{3\sqrt{3}}\right[$.

First of all, using a development up to order 2 of the equation (4.86) in $h = -\dfrac{2}{3\sqrt{3}}$, we obtain that $B''\left(-\dfrac{2}{3\sqrt{3}}\right) = -\dfrac{55}{2304}\sqrt{3} < 0$.

Let us suppose for a moment that $B''(h)$ would have a zero on $\left[-\dfrac{2}{3\sqrt{3}}, \dfrac{2}{3\sqrt{3}}\right[$, and let $h_0 > -\dfrac{2}{3\sqrt{3}}$ be the minimum of points $B''(h_0) = 0$ and $B''(h) < 0$, for all $h \in \left[-\dfrac{2}{3\sqrt{3}}, h_0\right[$.

Let us consider D, the tangent to Γ at the point $m_0 = (h_0, B(h_0))$. As $B''(h_0) = 0$, the order of contact between D and Γ is at least 2. Let v be a vector orthogonal to D, and $D(u)$ a linear parametrization of D. The function $\psi(u) = \langle Z(D(u)), v \rangle$ ($\langle.,.\rangle$ denoting the euclidean scalar product on $\mathbb{R}^2$) has a zero of order at least 1 in u_0, with $D(u_0) = m_0$. As $B''(h) < 0$ for all $h \in \left[-\dfrac{2}{3\sqrt{3}}, h_0\right[$, the corresponding arc of Γ is situated below D. The line D hence cuts $\Delta_0 = \left\{h = -\dfrac{2}{3\sqrt{3}}\right\}$ at a point n_0 above α_0. At this point, Z is directed downwards. On the other hand, in the points of D with abscissa $< h_0$ but near h_0, Z is directed towards the half plane above D. From this it follows that the function $\psi(u)$ must have a zero at some $u_1 \neq u_0$ with $D(u_1) \in]n_0, m_0[$.

However, the vector field Z being quadratic, the function $\psi(u)$ is polynomial of second degree in u; the existence of a double zero at u_0 and another zero u_1 implies then $\psi \equiv 0$ and hence that Γ *is a line segment*. This is of course not compatible with $B''\left(-\dfrac{2}{3\sqrt{3}}\right) < 0$, ending the proof of the theorem (see Figure 4.4). $\square$

To finish this section, we give a very useful algorithm due to Petrov [P] to obtain a bound for the number of zeros of any algebraic integral $I = \displaystyle\int_{\gamma_h} \omega$ for the cubic Hamiltonian $H(x,y) = y^2 - x + x^3$. From point i) above, we know that there exist polynomials P, Q in h such that

$$I = P(h)I_0(h) + Q(h)I_1(h).$$

We can find in [P] an algorithm to compute P, Q and also an estimate of the degrees of P, Q in term of the degree of ω. The number of zeros of $I(h)$ on $\left]-\dfrac{2}{3\sqrt{3}}, \dfrac{2}{3\sqrt{3}}\right[$ is the same number as for the function $G = P + QB$.

If K is the greatest common divisor of P and Q, we can write

$$P = KP_0, \quad Q = KQ_0, \quad P_0, Q_0 \text{ without common roots.}$$

The number of zeros for G is equal to the number of zeros of K plus the number for $G_0 = P_0 + Q_0 B$.

But, as P_0, Q_0 have no common roots, the number of zeros of G_0 is the same as the number of zeros of $g = B + \dfrac{P_0}{Q_0}$.

The function g is a solution of the Ricatti equation

$$9\left(\frac{4}{27} - h^2\right)g' = R_2\, g^2 + R_1\, g + R_0,\tag{4.87}$$

with

$$R_0 = \frac{N}{Q_0^2}\quad\text{and}$$

$$N = 9\left(\frac{4}{27} - h^2\right)(P_0'\, Q_0 - P_0\, Q_0') - 7Q_0^2 + 3hP_0\, Q_0 + \frac{5}{3}\, Q_0^2.\tag{4.88}$$

The crucial point is the following result of Petrov, based on Khovanskii's ideas:

Lemma 17 *Let $\alpha < \beta$, α, $\beta \in \left]-\dfrac{2}{3\sqrt{3}}\,,\,\dfrac{2}{3\sqrt{3}}\right[$, be two consecutive roots of Q_0. Then between two consecutive roots of g in $]\alpha,\beta[$, there exists at least one root of N.*

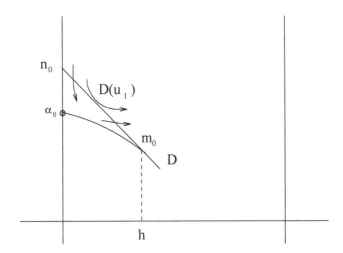

Figure 4.4

Proof. Let h_1, h_2 be two consecutive roots of g. If g' is also zero at any of these roots, it is also the case for R_0, following (4.87). Suppose that g' is not zero at h_1 and h_2.

Then $g'(h_1).g'(h_2) < 0$, and following (4.87), it is the same for R_0 and also for N. Then N has at least one root between h_1 and h_2. $\qquad\square$

An easy consequence of Lemma 17 (extended to the case of a multiple root of g) is that the number of roots of g between α and β, counted with multiplicity, is less than the number of roots for N between the same points, increased by one.

It follows that the total number of roots for g, with multiplicity, between $-\dfrac{2}{3\sqrt{3}}, \dfrac{2}{3\sqrt{3}}$, is bounded by $\deg(N) + \deg(Q_0) + 2$. From (4.88),

$$\deg(N) \leq 2\text{Sup} \left(\deg(P_0), \deg(Q_0)\right) + 1$$

and so the number of roots of I is bounded by the above bound plus the number of roots of K. Finally, we obtain

Corollary 2 *Suppose that* $I(h) = P(h)I_0(h) + Q(h)I_1(h)$, P, Q *are polynomials in* h. *Then the number of roots of* I, *with multiplicity, between* $-\dfrac{2}{3\sqrt{3}}$ *and* $\dfrac{2}{3\sqrt{3}}$, *is bounded by* $2 \, Sup \, \left(\deg(P), \deg(Q)\right) + 1$.

Example.
Let us consider the family of Section 3.5.3. As a consequence of the results in this section, the cyclicity may be computed as the supremum of the cyclicity for all subfamilies $\lambda(\varepsilon) = \left\{ f\{\lambda_1 = \varepsilon\overline{\lambda}_1, \, \lambda_5 = \varepsilon\overline{\lambda}_5, \, \lambda_2 = \varepsilon\overline{\lambda}_2, \, \lambda_4 = \varepsilon\overline{\lambda}_4 \mid (\overline{\lambda}_1, \overline{\lambda}_2, \overline{\lambda}_3, \overline{\lambda}_4, \overline{\lambda}_5) \in S^3 \right\}$.

For such a family

$$\delta(h, \lambda(\varepsilon)) = \varepsilon M_1(h, \overline{\lambda}) + 0(\varepsilon),$$

with

$$M_1(h, \overline{\lambda}) = (\overline{\varphi}_1 + \overline{\varphi}_2 \, \overline{\varphi}_4 \, h)I_0(h) + (\overline{\varphi}_5 - 3\overline{\varphi}_2 \, \overline{\varphi}_4)I_1(h). \qquad (4.89)$$

As a consequence of Corollary 2, we have that the total number of zeros of M, counted with multiplicity, at each $h_0 \in \left[-\dfrac{2}{3\sqrt{3}}, \dfrac{2}{3\sqrt{3}}\right[$, is less than 3. We have proved above that this multiplicity is in fact less than 2.

Chapter 5
Bifurcations of Elementary Graphics

After the regular limit periodic sets, the simplest limit periodic sets are the *elementary graphics*.

As was defined in Chapter 2, an elementary graphic for X_{λ_0} is an invariant immersion of S^1, made of a finite number of regular orbits and elementary (i.e., hyperbolic or isolated semi-hyperbolic) singular points. Limit sets of each regular orbit are contained in the set of singular points and the immersion is oriented by the orbit orientation.

Such an elementary graphic Γ may be monodromic, that is having a return map defined on some interval $[a, b[$ with $a \in \Gamma$. In this case we often call it a *polycycle*.

In this chapter we will deal with monodromic graphics with hyperbolic singular points and will call them either hyperbolic graphics or polycycles. The same methods apply also to the study of the non-monodromic case.

The simplest case corresponds to a graphic with only one hyperbolic saddle. We will call it a *saddle connection, or homoclinic loop*. The first two sections are devoted to their study.

The most important fact is that the Poincaré map defined along Γ is not differentiable at points whose ω-limit is one of the singular points because the transition map near an elementary point is not differentiable.

In the first section we will establish an expansion of the transition near a hyperbolic saddle using a natural unfolding of the logarithm. We will apply this expansion to study unfoldings of the saddle connections of finite codimension in the second section, and of analytic infinite codimension in the following one. In the last section we will present some recent results concerning general elementary polycycles, due to Mourtada, El Morsalani, Ilyashenko, Yakovenko and others. Finally, we will point out some open questions.

91

R. Roussarie, *Bifurcations of Planar Vector Fields and Hilbert's Sixteenth Problem*, Modern Birkhäuser Classics, DOI: 10.1007/978-3-0348-0718-0_5, © Springer Basel 1998

5.1   Transition map near a hyperbolic saddle point

Consider a C^∞ unfolding (X_λ) at a hyperbolic saddle point s_{λ_0} of the vector field X_{λ_0} where λ belongs to the parameter space $P \simeq \mathbb{R}^n$. Here we are only interested in the germ of a family at the point $(s_{\lambda_0}, \lambda_0)$. Therefore, without loss of generality, we can suppose that the family of vector fields (X_λ) is defined in a neighborhood V of $s_{\lambda_0} = s = (0,0) \in \mathbb{R}^2$, for parameter values λ in a neighborhood W of the origin in $\mathbb{R}^n$, and has a hyperbolic saddle at s for all $\lambda \in W$. We can also suppose that the local unstable and stable manifolds are given by $W^\nu = 0x \cap V$ and $W^s = 0y \cap V$. Finally, we suppose that s is the unique singular point of X_λ in V.

5.1.1   Normal form of X_λ near the saddle point

Let the eigenvalues of X_λ at s be given by $\lambda_2(\lambda)$, $\lambda_1(\lambda)$ with $\lambda_2(\lambda) < 0 < \lambda_1(\lambda)$, for $\lambda \in W$. Let $r(\lambda) = -\dfrac{\lambda_2(\lambda)}{\lambda_1(\lambda)}$. We call it the *ratio of hyperbolicity of X_λ at s*. Dividing X_λ by $\lambda_1(\lambda)$, we can assume that the eigenvalues are 1, $-r(\lambda)$ and that the 1-jet of X_λ at s is equal to

$$j^1 \, X_\lambda(0) = x \, \frac{\partial}{\partial x} \, - r(\lambda)y \, \frac{\partial}{\partial y}. \tag{5.1}$$

A first consequence of the hyperbolicity of s is the following result of finite determinacy:

Proposition 8 [Bon] *There exists a function $K(k) : \mathbb{N} \to \mathbb{N}$ such that $K(k) \to \infty$ for $k \to \infty$, and such that if Y_λ is any germ of C^∞ family of vector fields along $\{s\} \times W$ with the property*

$$j^{K(k)} \, (Y_\lambda - X_\lambda)(0) = 0, \tag{5.2}$$

then, the two family germs X_λ and Y_λ are C^k-conjugate. (This means that there exists a C^k germ of a family of diffeomorphisms g_λ defined on a neighborhood of $\{s\} \times W_1$, such that $(g_\lambda)_*(Y_\lambda) = X_\lambda$ on this neighborhood.)

Remark 22 *The result in [Bon] is proved for families of vector fields on $\mathbb{R}^p$, and gives an explicit function $K(k)$ which depends only on $j^1 \, X_\lambda(0)$. It is important to notice that this result does not depend on the possible resonances.*

Proposition 8 allows us to replace X_λ by a polynomial family, up to a C^k conjugacy. We now want to prove a version of the Dulac-Poincaré normal form theorem for the family of vector fields X_λ.

Proposition 9 *Let X_λ be a C^∞ family as above.*

1) Suppose that $r(\lambda_0) \notin \mathbb{Q}$. Then there exists a sequence of neighborhoods W_i of λ_0 in W, $i \geq 1$: $\lambda_0 \in \cdots W_{i+1} \subset W_i \subset \cdots \subset W_1$, such that for any $N \in \mathbb{N}$ and $\lambda \in W_{N+1}$,

$$j^{N+1} \, X_\lambda(s) \sim x \, \frac{\partial}{\partial x} \, + r(\lambda)y \, \frac{\partial}{\partial y}. \tag{5.3}$$

2) Suppose that $r(\lambda_0) = \dfrac{p}{q}$, p and q without common factors. Then there exist a sequence of neighborhoods as above and a sequence of smooth functions $\alpha_i(\lambda) : W_i \to \mathbb{R}$, $\alpha_1(\lambda) = p - qr(\lambda)$ on W_1, such that for any $N \in \mathbb{N}$ and $\lambda \in W_{N+1}$,

$$j^{(p+q)N+1} X_\lambda(s) \sim x \frac{\partial}{\partial x} + \left(-r(\lambda) + \frac{1}{q} \sum_{i=1}^{N} \alpha_{i+1}\,(\lambda)(x^p\ y^q)^i \right) y \frac{\partial}{\partial y}$$

$$= x \frac{\partial}{\partial x} + \frac{1}{q} \left(-p + \sum_{i=0}^{N} \alpha_{i+1}\,(\lambda)(x^p\ y^q)^i \right) y \frac{\partial}{\partial y}. \tag{5.4}$$

Here the sign $\sim$ denotes equivalence of jets. Formulas (5.3), (5.4) are equivalent to the following statement: X_λ is C^∞ equivalent to $X_\lambda^N + P_\lambda^N$ where X_λ^N is the right-hand polynomial family of vector field of (5.3) or (5.4) and P_λ^N is a C^∞ family on $V_1 \times W_{N+1}$ with respectively a $(N+1)$ or a $((p+q)N+1)$-jet at s equal to zero, for any $\lambda \in W_{N+1}$.

Proof. The proof given in [R4] for the resonant case $p = q = 1$ is easily extended. It is based on the following remarks. There are no resonances if $r(\lambda_0) \notin \mathbb{Q}$. All the resonance relations $\lambda_i - \sum_{j=1}^{2} n_j\,\lambda_j = 0$, $i = 1, 2$, are generated by the unique relation

$$p\lambda_1(\lambda_0) + q\lambda_2(\lambda_0) = 0$$

if $r(\lambda_0) = p/q$. By continuity, for each N we can find a neighborhood W_N of λ_0 in W_1 such that this remains valid for all $\lambda \in W_N$ (of course, the neighborhoods W_N form a decreasing sequence). We construct the normal form up to order $(p+q)N+1$, using subspaces of homogeneous vector fields, which are independent of $\lambda \in W_N$. $\qquad\square$

Combining the two above propositions, we see that, at each order of differentiability, we can replace the given family near the saddle point by a polynomial one.

Theorem 13 *Let X_λ be a C^∞ family, as defined above, near a saddle point $s_{\lambda_0} = s$ of X_{λ_0}. There exists a function $N(k) : \mathbb{N} \to \mathbb{N}$ such that in some neighborhood of s and for $\lambda \in W_{N(k)}$, the family X_λ is C^k-equivalent to the polynomial family*

$$x \frac{\partial}{\partial x} + \left(r(\lambda) + \frac{1}{q} \sum_{i=1}^{N(k)} \alpha_{i+1}\,(\lambda)(x^p\ y^q)^i \right) y \frac{\partial}{\partial y} \tag{5.5}$$

if $r(\lambda_0) = p/q$. If $r(\lambda_0) \notin \mathbb{Q}$, all the $\alpha_{i+1}(\lambda) \equiv 0$ for $i \geq 1$.

Proof. It suffices to take $N(k)$ such that $(p+q)N(k)+1 > K(k)$ in case of resonance p/q and $N(k)+1 > K(k)$ when $r(\lambda_0) \notin \mathbb{Q}$ and to apply the above two propositions. $\qquad\square$

Remark 23 *The function $\alpha_1(\lambda)$ is uniquely defined by the relation $\alpha_1(\lambda) = p - qr(\lambda)$. It is not the same for the other resonant quantities $\alpha_i(\lambda)$, $i \geq 2$, in the resonant case $r(\lambda_0) = p/q$. Nevertheless, we have seen that we can choose smooth α_i if X_λ is C^∞, and clearly analytic α_i if X_λ is an analytic family.*

5.1.2 The structure of the transition map for the normal family

If $r(\lambda_0) \notin \mathbb{Q}$, then we have seen that X_λ is C^k equivalent to the linear family $X_\lambda^N = x \dfrac{\partial}{\partial x} - r(\lambda) \, y \, \dfrac{\partial}{\partial y}$, if one restricts λ to $W_{N(k)+1}$. The transition map for this linear family between $\sigma = [0,1[\times \{1\}$ and $\tau = \{1\} \times] - 1, 1[$ is $x \to x^{r(\lambda)}$.

If now $r(\lambda_0) = p/q$, we have seen that X_λ factorizes up to a C^k equivalence, through a polynomial family of special type.

More generally, we will consider the *analytic normal family* X_α^N,

$$X_\alpha^N = x \frac{\partial}{\partial x} + \frac{1}{q} \left(-p + \sum_{i=0}^\infty \alpha_{i+1} \, (x^p \, y^q)^i \right) y \, \frac{\partial}{\partial y}, \tag{5.6}$$

where $P_\alpha(u) = \displaystyle\sum_{i=0}^\infty \alpha_{i+1} \, u^{i+1}$ is an analytic entire function of $u \in \mathbb{R}$, and $\alpha = (\alpha_1, \alpha_2, \ldots) \in A$, where A is the set

$$A = \{\alpha = (\alpha_1, \alpha_2, \ldots) \mid \; \mid \alpha_1 \mid < \frac{1}{2} \; , \; \mid \alpha_i \mid < M \; \text{ for } i \geq 2\} \tag{5.7}$$

and $M > 0$ is a fixed constant.

The axes $0x$ and $0y$ are the unstable and stable manifolds of X_α^N. If σ, τ are transversal segments as above, then the flow of X_α^N defines a transition map $D_\alpha(x)$ from σ to τ, which extends continuously to $D_\alpha(0) = 0$ (see Figure (5.1)). This map is analytic in (x, α) for $x \neq 0$. We want to study its properties at $\{x = 0\}$.

Making the singular change of variables $u = x^p \, y^q$, $x = x$, the differential equation of X_α^N is brought in the following form:

$$\begin{cases} \dot{x} &= x \\ \dot{u} &= P_\alpha(u) = \displaystyle\sum_{i=1}^\infty \alpha_i \, u^i. \end{cases} \tag{5.8}$$

We see that the variables are separated in (5.8). The first equation gives $x(t, x) = x e^t$. Let us look at the second one; $P_\alpha(u)$ is analytic for $\mid u \mid \leq 1$, and $\alpha \in A$. Call $u(t, u)$ the solution verifying the initial condition $u(0, u) = u$.

For each t we can expand it in series in u,

$$u(t, u) = \sum_{i=1}^\infty g_i(t) u^i. \tag{5.9}$$

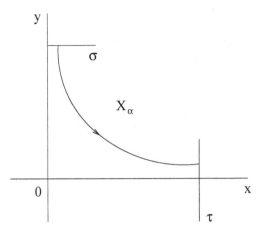

Figure 5.1

We have $g_1(t) = e^{\alpha_1 t}$ and $g_i(0) = 0$ for $i \geq 2$.

We want to study the form of the functions g_i and the convergence of the above series in function of t. For this, we will compare $u(t, u)$ with the solution of the hyperbolic equation

$$\dot{U} = \frac{1}{2} U + M \sum_{i=1}^{\infty} U^{i+1}. \tag{5.10}$$

The following estimations hold:

Lemma 18 *Let* $U(t, u) = \displaystyle\sum_{i=1}^{\infty} G_i(t) u^i$ *be the power series expansion of the trajectories of (5.10). Then, for each* $i \geq 1$ *and* $t \geq 0$,

$$\mid g_i(t) \mid \leq G_i(t) \quad \text{for any } \alpha \in A.$$

Proof. Substituting (5.9) in the equation $\dfrac{\partial u}{\partial t}(t, u) = P_\alpha(u(t, u))$, we obtain the system E_g of equations for the functions $g_i(t)$:

$$
\begin{aligned}
\dot{g}_1(t) &= \alpha_1\, g_1 \\
\dot{g}_2(t) &= \alpha_1\, g_2 + \alpha_2\, g_1^2 \\
\dot{g}_3(t) &= \alpha_1\, g_3 + 2\alpha_2\, g_1\, g_2 + \alpha_3\, g_2^2.
\end{aligned}
$$

More generally,

$$\dot{g}_i = \alpha_1\, g_i + P_i(\alpha_2, \dots, \alpha_i\,;\, g_1, \dots, g_{i-1}) \quad \text{for } i \geq 2$$

where P_i is a polynomial in $\alpha_2, \ldots, \alpha_i, g_1, \ldots, g_{i-1}$ with positive rational coefficients.

Now, let $U(t, u)$ be the trajectory of $\dot{U} = P_\alpha(U)$ with $\alpha = \left(\frac{1}{2}, M, M, \ldots \right)$. The functions $G_i(t)$ verify the system E_G:

$$\dot{G}_1 = \frac{1}{2} G_1$$
$$\dot{G}_2 = \frac{1}{2} G_2 + M G_1^2,$$

and more generally,

$$\dot{G}_i = \frac{1}{2} G_i + P_i(M, \ldots, M \; ; \; G_1, \ldots, G_{i-1})$$

(with the same polynomials P_i as above).

We can solve the system E_G by

$$G_1(t) = e^{1/2t} , \; G_2(t) = \psi_2(t) e^{\frac{1}{2}t} \; \text{with} \; \psi_2(t) = \int_0^t e^{-\frac{1}{2}\tau} M \, G_1^2(\tau) d\tau,$$

and more generally,
$$G_i(t) = \psi_i(t) e^{\frac{1}{2}t} \; \text{with}$$

$$\psi_i(t) = \int_0^t e^{-\frac{1}{2}\tau} P_i\Big(M, \ldots, M \; ; G_1(\tau), \ldots, G_{i-1}(\tau)\Big) d\tau. \qquad (5.11)$$

It follows easily from these formulas that $G_i(t) > 0$ for $t > 0$. We now show that

$$\mid g_i(t) \mid \leq G_i(t)$$

for each $t \geq 0$. First, this is trivially true for $i = 1$:

$$\mid g_1(t) \mid \leq e^{\mid \alpha_1 \mid t} \; \leq e^{\frac{1}{2}t} \; = G_1(t). \qquad (5.12)$$

Suppose that we have shown that $\mid g_j(t) \mid \leq G_j(t)$ for each $j : 1 \leq j \leq i - 1$, and any $t \geq 0$. We compare the two equations

$$\dot{g}_i(t) = \alpha_1 \, g_i + P_i \, (\alpha_2, \ldots, \alpha_i \; ; \; g_1, \ldots, g_{i-1})$$

and

$$\dot{G}_i(t) = \frac{1}{2} G_i + P_i \, (M, \ldots, M \; ; \; G_1, \ldots, G_{i-1}).$$

As the coefficients in P_i are positive, we have

$$\mid P_i(\alpha_2, \ldots, \alpha_i \; ; \; g_1, \ldots, g_{i-1}) \mid \leq P_i(\mid \alpha_2 \mid, \ldots, \mid \alpha_i \mid \; ; \; \mid g_1 \mid, \ldots, \mid g_{i-1} \mid)$$

$$\leq \; P_i(M, \ldots, M \; ; \; G_1, \ldots, G_{i-1}). \qquad (5.13)$$

Now, $G_1(0) = 1$ and $G_i(0) = 0$ for $i \geq 2$. Hence we have $\dot{G}_i(0) = P_i$ $(M, \ldots, M; G_1(0), \ldots, G_{i-1}(0)) = MG_1(0)^i = M$ and also $| \dot{g}_i(0) | \leq | \alpha_i |$ $| g_1(0) |^i \leq | \alpha_i | < M$. Therefore we have $g_i(0) = G_i(0) = 0$ and $| \dot{g}_i(0) | < \dot{G}_i(0)$. By continuity, this gives $| \dot{g}_i(t) | < \dot{G}_i(t)$, for t small enough.

We show now that this inequality holds for any $t \geq 0$ (and so we will have $| g_i(t) | \leq G_i(t)$ for any $t \geq 0$).

Suppose on the contrary that $t_0 > 0$ is the inferior bound of the values of t for which $| \dot{g}_i(t) | \geq \dot{G}_i(t)$. We have, for all $t \in [0, t_0]$, $| \dot{g}_i(t) | \leq \dot{G}_i(t)$, so we also have $| g_i(t) | \leq G_i(t)$ for all $t \in [0, t_0]$. Now, for $t = t_0$,

$$\dot{g}_i(t_0) = \alpha_1 g_i(t_0) + P_i(\alpha_2, \ldots, \alpha_i ; g_1(t_0), \ldots, g_{i-1}(t_0))$$
$$\dot{G}_i(t_0) = \frac{1}{2} G_i(t_0) + P_i(M, \ldots, M ; G_1(t_0), \ldots, G_{i-1}(t_0)).$$

By induction on i, we know that $G_j(t_0) \geq g_j(t_0)$ for $2 \leq j \leq i - 1$. By the choice of t_0, we already know that $G_i(t_0) \geq | g_i(t_0) |$, so the inequality $| \alpha_1 | < \frac{1}{2}$ implies that $| \dot{g}_i(t_0) | < \dot{G}_i(t_0)$. But, by continuity, this strict inequality is valid for $t > t_0$, t near t_0; this last point contradicts the definition of t_0. □

Next, we prove the following:

Lemma 19 *There exist constants C, $C_0 > 0$ such that*

$$| g_i(t) | \leq C_0 [Ce^{t/2}]^i \quad \text{for any } i \geq 1, \ t \geq 0 \quad \text{and any } \alpha \in A.$$

Proof. Using Lemma 18, it suffices to show that $G_i(t) \leq C_0 [Ce^{\frac{t}{2}}]^i$ for some constants C_0, C, and $i \geq 1$, $t \geq 0$, $\alpha \in A$. Recall that the function $U(t, u) = \sum_{i \geq 1} G_i(t) u^i$ is a trajectory of the hyperbolic 1-dimensional vector field $X = P(u) \frac{\partial}{\partial u}$ with $P(u) = \frac{1}{2} u + M \sum_{i=1}^{\infty} u^{i+1}$. From the analytic linearization theorem of Poincaré, there exists an analytic diffeomorphism $g(u) = u + O(u)$, converging for $| u | \leq K_1$, K_1 fixed, such that

$$g_* \left(P(u) \frac{\partial}{\partial u} \right) = \frac{1}{2} u \frac{\partial}{\partial u}. \tag{5.14}$$

This diffeomorphism g sends the flow $U(t, u)$ of $P \frac{\partial}{\partial u}$ into $U_0(t, u) = ue^{\frac{1}{2}t}$, the flow of $\frac{1}{2} u \frac{\partial}{\partial u}$.

This means that $U_0(t, g(u)) = g \circ U(t, u)$ for $| u |, | U(t, u) | \leq K_1$.

As g is invertible for $| u | \leq K_1$, there exist constants b, B, $0 < b < B$ such that

$$b | u | \leq | g(u) | \leq B | u | \quad \text{for } | u | \leq K_1. \tag{5.15}$$

Suppose that $\mid u \mid \leq \dfrac{b}{B} \, K_1 \, e^{-\frac{1}{2}t}$. Then

$$\mid g(u) \mid \leq B \mid u \mid \leq b \, K_1 \, e^{-\frac{1}{2}t}, \qquad (5.16)$$

and $\mid U_0(t, g(u)) \mid = \mid g(u) \mid e^{\frac{t}{2}} \leq bK_1$.

Now, we have $U(t, u) = g^{-1} \circ U_0(t, g(u))$.

This implies that

$$\mid U(t, u) \mid \leq \frac{1}{b} \mid U_0(t, g(u)) \mid \leq K_1. \qquad (5.17)$$

Using Cauchy's inequalities for the coefficients $G_i(t)$, we find

$$\mid G_i(t) \mid \leq \frac{\mathrm{Sup}\ \{\mid U(t, u) \mid\ \mid\ \mid u \mid = R(t)\}}{(R(t))^i} \leq \frac{K_1}{(R(t))^i} \qquad (5.18)$$

where $R(t) = \dfrac{b}{B} \, K_1 \, e^{-\frac{t}{2}}$.

Finally, we obtain

$$\mid G_i(t) \mid \leq K_1 \left(\frac{B}{b} \, K_1^{-1}\right)^i \, e^{i\frac{t}{2}},$$

which gives the desired estimate, with $C_0 = K_1$ and $C = \dfrac{B}{b} \, K_1^{-1}$. $\qquad\square$

We show below that the functions g_i are analytic for $t > 0$. For the moment, we note that the formula $\dfrac{\partial u}{\partial t} (t, u) = P_\alpha(u(t, u))$ shows that the series in u of $\dfrac{\partial u}{\partial t}$ has the same radius of convergence as $u(t, u)$. (Recall that $P_\alpha(u)$ is supposed to be an entire function.) The same is true for any derivative $\dfrac{\partial^k u}{\partial t^k} (t, u)$, by induction on k. This remark gives an estimate for the coefficients $\dfrac{d^k g_i}{dt^k} (t)$ of this derivative.

Using Cauchy's inequalities as above,

$$\left| \frac{d^k g_i(t)}{dt^k} \right| \leq \frac{\mathrm{Sup}\ \{\mid \dfrac{\partial^k u}{\partial t^k} (t, u) \mid;\ \mid u \mid = R(t)\}}{\mid R(t) \mid^i},$$

which gives $\left| \dfrac{d^k g_i}{dt^k} (t) \right| \leq C_k \, (C e^{\frac{t}{2}})^i$ for some $C_k > 0$.

We have proved

Lemma 20 *For each $k \geq 1$, there exists a constant $C_k > 0$ such that*

$$\left| \frac{d^k g_i}{dt^k} (t) \right| \leq C_k \, (C.e^{\frac{t}{2}})^i \quad \text{for any } i \geq 1,\ t \geq 0,\ \alpha \in A \qquad (5.19)$$

(C is the same as in Lemma 19).

We now establish more precisely the form of the functions $g_i(t)$. For this, we introduce the function

$$\Omega(\alpha_1, t) = \frac{e^{\alpha_1 t} - 1}{\alpha_1} \quad \text{for } \alpha_1 \neq 0 \text{ and}$$

$$\Omega(0, t) = t.$$

With this notation we have

Proposition 10 *For each $k \geq 1$, $g_k(t) = e^{\alpha_1} Q_k(t)$, where Q_k is a polynomial of degree $\leq k - 1$ in Ω. The coefficients of Q_k are polynomial in $\alpha_1, \ldots, \alpha_k$. More precisely,*

$$Q_k = \alpha_k \Omega + \bar{Q}_k(\alpha_1, \ldots, \alpha_k, \Omega), \tag{5.20}$$

where $\bar{Q}_k$ is a polynomial of degree $\leq k - 1$ in Ω, with coefficients in $\mathcal{I}(\alpha_1, \ldots, \alpha_{k-1}) \cap \mathcal{I}(\alpha_1, \ldots, \alpha_k)^2 \subset \mathbb{Z}[\alpha_1, \ldots, \alpha_k]$. Here, $\mathcal{I}(u, v, \ldots)$ stands for the polynomial ideal generated by $u, v, \ldots$.

Proof. Rewrite the system E_g for the g_i,

$$\dot{g}_1 = \alpha_1 g_1$$
$$\dot{g}_2 = \alpha_1 g_2 + \alpha_2 g_1^2$$
$$\vdots$$
$$\dot{g}_k = \alpha_1 g_k + P_k(\alpha_2, \ldots, \alpha_k ; g_1, \ldots, g_{k-1}).$$

The polynomial P_k is obtained from the coefficient of u^k in the expansion $\sum_{j \geq 2} \alpha_j \left[\sum_{i \geq 1} g_i u^i \right]^j$. It follows easily that P_k is homogeneous linear in $\alpha_2, \ldots, \alpha_k$.

Each polynomial

$$g_1^{\ell_1} \cdots g_{k-1}^{\ell_{k-1}} \quad \text{is such that} \quad \sum_{j=1}^{k-1} \ell_j \geq 2 \quad \text{and} \quad \sum_{j=1}^{k-1} j.\ell_j = k. \tag{5.21}$$

First, we show that $g_k(t) = e^{\alpha_1 t} Q_k(t)$ with Q_k a polynomial of degree $\leq k-1$ in Ω, and polynomial coefficients in $\alpha_1, \ldots, \alpha_k$ (in particular: $g_1(t) = e^{\alpha_1 t}$, $g_2(t) = \alpha_2 e^{\alpha_1 t} \Omega, \ldots$).

Consider the equation for g_k,

$$\dot{g}_k = \alpha_1 g_k + P_k(\alpha_2, \ldots, \alpha_k ; g_1, \ldots, g_{k-1}).$$

We use induction on k, so let us suppose that for each $j \leq k - 1$, $g_j(t) = e^{\alpha_1 t} Q_j(t)$, as above. Notice that $e^{\alpha_1 t} = \alpha_1 \Omega + 1$, so each g_j has degree $\leq j$ in Ω. It now follows from (5.23) that

$$P_k(\alpha_2, \ldots, \alpha_k ; g_1, \ldots, g_{k-1}) = e^{2\alpha_1 t} X_k(\Omega),$$

where X_k is a polynomial of degree $\leq k - 2$ in Ω (to see this point, replace in each monomial $g_1^{\ell_1} \cdots g_{k-1}^{\ell_{k-1}}$ of P_k, a product of two factors g_i, g_j by $e^{2\alpha_1 t} Q_i Q_j$ and the other factors g_ℓ by $(\alpha_1 \Omega + 1) Q_\ell$).

Now, $g_k = e^{\alpha_1 t} Q_k$, with

$$Q_k(t) = \int_0^t e^{-\alpha_1 \tau} P_k (\alpha_2, \ldots, \alpha_k \, ; \, g_1, \ldots, g_{k-1}) d\tau$$

$$Q_k(t) = \int_0^t e^{\alpha_1 \tau} X_k (\Omega) d\tau = \int_0^t X_k (\Omega) \dot{\Omega} d\tau \qquad (5.22)$$

(because $\dot{\Omega} = e^{\alpha_1 t}$).

It follows from (5.22) that $Q_k(t)$ is a polynomial of degree $\leq k - 1$ in Ω. From the induction, it follows easily that the coefficients are polynomials in $\alpha_1, \ldots, \alpha_k$. To obtain the precise form of the statement, notice that

$$P_k(\alpha_2, \ldots, \alpha_k \, ; \, g_1, \ldots, g_{k-1}) = \alpha_k \, g_1^k + \widetilde{P}_k,$$

for $k \geq 2$, where $\widetilde{P}_k$ is linear in $\alpha_2, \ldots, \alpha_{k-1}$ and each monomial in $\widetilde{P}_k$ contains at least one of the g_i with $i \geq 2$. Moreover, we know that the coefficients of such a g_i are in $\mathcal{I}(\alpha_1, \ldots, \alpha_i)$. So, the coefficients of $\widetilde{P}_k$ are in $\mathcal{I}(\alpha_1, \ldots, \alpha_{k-1}) \cap \mathcal{I}(\alpha_2, \ldots, \alpha_k)^2$.

Now,

$$Q_k = \alpha_k \int_0^t e^{(k-1)\alpha_1 \tau} \, d\tau + \int_0^t e^{-\alpha_1 \tau} \, \widetilde{P}_k(\tau) d\tau. \qquad (5.23)$$

Consider the first term,

$$\int_0^t e^{(k-1)\alpha_1 \tau} \, d\tau = \frac{e^{(k-1)\alpha_1 t} - 1}{(k-1)\alpha_1}.$$

Using $e^{\alpha_1 t} = \alpha_1 \Omega(t) + 1$, we obtain

$$e^{(k-1)\alpha_1 t} = 1 + (k-1)\alpha_1 \, \Omega + \alpha_1^2 \, S(\Omega), \qquad (5.24)$$

where $S(\Omega)$ is a polynomial in Ω. Hence, we obtain

$$\alpha_k \int_0^t e^{(k-1)\alpha_1 \tau} \, d\tau = \alpha_k \, \Omega + \frac{\alpha_k \, \alpha_1}{k-1} \, S(\Omega). \qquad (5.25)$$

The term $\int_0^t e^{-\alpha_1 \tau} \, \widetilde{P}_k \, d\tau$ gives a polynomial in Ω, with coefficients in $\mathcal{I}(\alpha_1, \ldots, \alpha_{k-1}) \cap \mathcal{I}(\alpha_2, \ldots, \alpha_k)^2$, so we finally obtain $Q_k(t) = \alpha_k \, \Omega + \overline{Q}_k$ with $\overline{Q}_k$ as in the statement. $\qquad \square$

We now go back to the map $D_\alpha(x)$.

The time to go from σ to τ is equal to $t(x) = -Lnx$. Notice that $u_{|\sigma} = x^p$ and $u_{|\tau} = y^q$. Thus we can compute $D_\alpha(x)$ using $u(t, x)$,

$$D_\alpha(x)^q = u(-Lnx, \ x^p) \ , \text{ for } x > 0 \ , with \ D_\alpha(0) = 0. \tag{5.26}$$

It is not difficult to see that D_α is well defined for $x \in [0, X]$, where X is some value > 0, and is analytic in (x, α), for $x \neq 0$, $\alpha \in A$. We want to study its behavior at $x = 0$. For this, we note that Lemma 5 implies that the convergence radius of the series $\Sigma \ g_i(t)u^i$ is greater that $\dfrac{1}{C} \ e^{-\frac{1}{2}t}$, for each $t > 0$. Therefore, the series $\displaystyle\sum_i g_i(t)x^{p_i}$ converges for x small enough and each $t < -2Lnx$. In particular, the series converges for $t = -Lnx$. We can use this series to compute $D_\alpha(x)$,

$$D_\alpha(x)^q = \sum_{i=1}^{\infty} g_i(-Lnx)x^{pi}. \tag{5.27}$$

The convergence is normal on the interval $[0, X]$, for some $X > 0$. We can now use the estimates on g_i, $\dfrac{d^k g_i}{dt^k}$ obtained in Lemmas 19, 20 to prove the following:

Proposition 11 *For any $k \in \mathbb{N}$ there exists $K(k)$ such that*

$$D_\alpha(x)^q = \sum_{i=1}^{K(k)} g_i(-Lnx)x^{pi} + \psi_k, \tag{5.28}$$

where $\psi_k(x, \alpha)$ is a C^{kp} function in (x, α), kp-flat at $x = 0$ (i.e., $\psi_k(0, \alpha) = \dfrac{\partial \psi_k}{\partial x} (0, \alpha) = \cdots = \dfrac{\partial^{kp} \psi_k}{\partial x^{kp}} (0, \alpha) = 0$).

Proof. Given k, we want to find $K(k)$ such that

$$(D_\alpha^q)^K \ (x) = \sum_{K+1}^{\infty} g_i(-Lnx)x^i$$

is a C^{kp}, kp-flat function.

We show that the series $(D_\alpha^q)^K$ can be derived term by term. First, we have

$$\frac{d}{dx}[g_j(-Lnx)x^{pj}] = -g_j^{(1)}(-Lnx)x^{pj-1} + pjg_j(-Lnx)x^{pj-1} \tag{5.29}$$

$$\left(g_j^{(1)} = \frac{dg_j}{dx} \right).$$

Now, using the estimates of Lemma 20, we have

$$\left| g_j^{(1)}(-Lnx) \right| \le C_1 \mid Cx \mid^{-j/2},$$

and using Lemma 19,

$$\mid g_j(-Lnx) \mid \le C_0 \mid Cx \mid^{-j/2} .$$

Thus we have

$$\left| \frac{d}{dx} \left[g_j(-Lnx)x^{pj} \right] \right| \le j M_1 \mid Cx \mid^{(p-\frac{1}{2})j-1}, \tag{5.30}$$

for some constant $M_1 > 0$.

More generally, using Lemma 2 we have

$$\left| \frac{d^s}{dx^s} \left[g_j(-Lnx)x^{pj} \right] \right| \le \frac{j!}{(j-s)!} M_s \mid Cx \mid^{(p-\frac{1}{2})j-s}, \tag{5.31}$$

for each $s \le j$ and some constant $M_s > 0$.

It follows from these estimates that if $(p - \frac{1}{2})K > k$ and if $0 \le s \le k$, then the series

$$\sum_{j \ge K+1} \frac{d^s}{dx^s} \left(g_j(-kn(x)x^{pj} \right)$$

converges and is equal to 0 at $x = 0$, giving that the function $(D_\alpha^q)^K = \displaystyle\sum_{j \ge K+1} \cdots$

is k-flat and C^k.   □

We define the function $w(x, \alpha_1)$ by

$$w(x, \alpha_1) = \frac{x^{-\alpha_1} - 1}{\alpha_1} \quad \text{if } \alpha_1 \ne 0$$

$$w(x, 0) = -Lnx. \tag{5.32}$$

This function is related to the above function $\Omega(\alpha_1, t)$ by

$$w(x, \alpha_1) = \Omega(\alpha_1, -Lnx). \tag{5.33}$$

Note that for each $k > 0$, $x^k w \to -x^k \, Lnx$ as $\alpha_1 \to 0$ (uniformly for $x \in [0, X]$, for any fixed $X > 0$). We consider finite combinations of monomials $x^i \, w^j$ with $i, j \in \mathbb{N}$ and $0 \le j \le i$. These functions $x^i \, w^j$ form a totally ordered set with the following order:

$$x^i w^j \prec x^{i'} w^{j'}$$

$\Longleftrightarrow i' > i$ or $i = i'$ and $j > j'$. We have $1 \prec xw \prec x \prec x^2 w^2 \prec x^2 w \prec x^2 \prec \cdots$.

The notation $x^i w^j + \cdots$ means that after the sign $+$ we find a finite combination of $x^{i'} w^{j'}$ of strictly greater order. Then, for the transition map $D_\alpha(x)$, we have the following (x, w)-expansion of order k:

Theorem 14 *The transition map D_α of X^N_α defined by (5.28) (relative to the segments σ, τ defined above) has the following (x, ω)-expansion of order kp:*

$$(D_\alpha(x))^q = x^p + \alpha_1[x^p\omega + \cdots] + \alpha_2[x^{2p}\omega + \cdots] + \cdots$$

$$+\alpha_K[x^{Kp}\,\omega + \cdots] + \psi_K(x, \alpha), \qquad (5.34)$$

for any $k \in \mathbb{N}$.

The index $K(k)$ is defined in Proposition 11, and each term between the brackets is a finite combination of monomials $x^i\omega^j$ (with the above convention); the coefficients of the unwritten monomials $x^i\omega^j$ after the signs $+$ are polynomial functions of the α_s, $s = 1, \ldots, K$, which are zero if $\alpha = 0$. The remainder ψ_k is a C^{kp} function in (x, α), which is kp-flat for $x = 0$ and any α,

$$\left(\psi_k(0, \alpha) = \frac{\partial \psi_k}{\partial x}\,(0, \alpha) = \cdots = \frac{\partial^{kp}}{\partial x^{kp}}\,\psi_k(0, \alpha) = 0\right).$$

Proof. Proposition 10 gives

$$g_k(-Lnx) = e^{-\alpha_1 Lnx}\,Q_k(-Lnx)$$

$$= x^{-\alpha_1}\left(\alpha_k\omega + \bar{Q}_k(\alpha_1, \ldots, \alpha_k, \omega)\right), \qquad (5.35)$$

with $\bar{Q}_k$ of degree $\leq k-1$ in ω and coefficients in $\mathcal{I}(\alpha_1, \ldots, \alpha_{n-1}) \cap \mathcal{I}(\alpha_2, \ldots, \alpha_k)^2$. Hence, the general term $g_k(-Lnx)x^{pk}$ in $(D_\alpha(x))^q$ is equal to

$$g_k(-Lnx)x^{pk} = x^{pk-\alpha_1}\,(\alpha_k\,\omega + \bar{Q}_k). \qquad (5.36)$$

Using $x^{-\alpha_1} = \alpha_1\omega + 1$, this term can be rewritten as

$$g_k(-Lnx)x^{pk} = \alpha_k x^{pk}\omega + \alpha_1\alpha_k\,x^{pk}\,\omega^2 + (1 + \alpha_1\omega)x^{pk}\,\bar{Q}_k, \qquad (5.37)$$

for $k \geq 2$ and $g_1(-Lnx)x^p = x^{p-\alpha_1} = \alpha_1\,x^p\,\omega + x^p$. \qquad (5.38)

We thus have

$$(D_\alpha(x))^q = x^p + \alpha_1\,x^p\,\omega + \alpha_2\,x^{2p}\,\omega + \alpha_1\,\alpha_2\,x^{2p}\,\omega^2 + (1 + \alpha_1\omega)x^{2p}\,\bar{Q}$$

$$+\alpha_3\,x^{3p}\,\omega + \alpha_1\,\alpha_3\,x^{3p}\,\omega^2 + x^{3p}(1 + \alpha_1\,\omega)\bar{Q}_3 + \cdots + \psi_K, \qquad (5.39)$$

where $+\cdots$ denotes an expansion of $x^{ps}g_s(-Lnx)$ for $4 \leq s \leq K(k)$. We now can rearrange the sum $\displaystyle\sum_{i=1}^{K(k)} g_i(-Lnx)x^{pi}$ in the following way: first, we take all the terms whose coefficients are divisible by α_1; next, all the remainders (not divisible by α_1) but divisible by α_2 and so on, until α_K. We obtain the following expansion:

$$
\begin{aligned}
D_\alpha(x)^q = x^p \;\; & + \;\; \alpha_1[x^p\,\omega + \alpha_2\,x^{2p}\,\omega + x^{2p}\,\omega\overline{Q}_2 + \alpha_3\,x^{3p}\,\omega^2 + \cdots] \\
& + \;\; \alpha_2\,[x^{2p}\,\omega + \text{ terms in } x^{3p}\,\overline{Q}_3,\ldots,x^{Kp}\,\overline{Q}_K \\
& \qquad\qquad \text{divisible by } \alpha_2,\;\; \text{not by } \alpha_1] \\
& \;\;\vdots \\
& + \;\; \alpha_K\,x^{Kp}\,\omega + \psi_k(x,\alpha).
\end{aligned}
$$

Considering this expansion, it is clear that each term after $x^{sp}\,\omega$ in the bracket related to α_s is of order greater that $x^{sp}\,\omega$ and has a coefficient in $\mathcal{I}(\alpha_1,\ldots,\alpha_K)$ (because it comes from some coefficients in $\mathcal{I}(\alpha_1,\ldots,\alpha_K)^2$ divided by α_s). The sum from 1 to K contains all monomial terms in x,ω coming from the expansion $\sum_{i=1}^{K} g_i(-Lnx)x^{ip}$ and we know that the remaining term ψ_k obtained in Proposition 4 is $\mathcal{C}^k$, k-flat at $x = 0$. This completes the proof. $\square$

5.1.3 The structure of the transition map of X_λ

We now return to the initial $\mathcal{C}^\infty$ family X_λ. We assume to have chosen a fixed system of coordinates (x,y) for which the saddle point s_λ is at the origin, the $0x$ and $0y$ axes are local unstable and stable manifolds of X_λ respectively, for each $\lambda \in W_1$, and the 1-jet of X_λ is given by (5.1).

Now take transversal segments σ, τ to $0y$ and $0x$ respectively: σ is parametrized by $x \in [0,X]$ and τ by $y \in [-Y,Y]$ for some $X,Y > 0$. Theorem 13 gives a $\mathcal{C}^k$ equivalence of (X_λ) with a polynomial normal form family $(X^N_{\alpha(\lambda)})$, for any $k \in \mathbb{N}$. The family (X^N_α), for $\alpha(\lambda) = (\alpha_1(\lambda),\ldots,\alpha_N(\lambda))$, is defined in Section 1.1. This $\mathcal{C}^k$ equivalence defines $\mathcal{C}^k$ families of diffeomorphisms $\Phi_\lambda(x)$, $\psi_\lambda(y)$ on $\mathbb{R}$, in a neighborhood of $\Phi_\lambda(0) = \psi_\lambda(0) = 0$ such that if D_α is the transition map for X^N_α, then we have

$$D_\lambda(x) = \psi_\lambda \circ D_{\alpha(\lambda)} \circ \Phi_\lambda(x). \tag{5.40}$$

The family (X^N_α) is linear, if $r(\lambda_0) \notin \mathbb{Q}$ and is given by the polynomial normal form (Proposition 9) if $r(\lambda_0) = \dfrac{p}{q}$. In this last case, we can apply to it the results of Section 1.2, in particular Theorem 14. We have an (ω,x)-expansion at order k for any k, which depends only on $\alpha_1(\lambda),\ldots,\alpha_N(\lambda)$ because all the $\alpha_i(\lambda) \equiv 0$, for $i \geq N+1$:

$$(D_{\alpha(\lambda)}\,(x))^q = x^p + \alpha_1(\lambda)[x^p\omega + \cdots] + \cdots$$
$$+\alpha_N(\lambda)[x^{Np}\,\omega + \cdots] + \psi_k\,(x,\lambda), \tag{5.41}$$

as in Theorem 14. The resonant coefficients $\alpha_i(\lambda)$ are independent of k, but such is not the case for the expansions in the brackets.

5.1.3.1 Dulac Series for D_{λ_0}

In this subsection, we verify that the transition map near an hyperbolic saddle is quasi-regular (we have used this fact in Chapter 3). In order to expand the transition $D(x)$ for the saddle point of a vector field X, we can use formulas (5.40) and (5.41) for a trivial family (X_λ constant and equal to X and $D_{\lambda_0} = D$). Then $\alpha_1 = 0$ and $\omega = -Ln\ x$.

If $r = r(0) \notin Q$ we have $D(x) = \psi \circ D_0 \circ \varphi(x)$ for ψ, φ $\mathcal{C}^k$ diffeomorphisms and $D_0(x) = x^r$.

If $r = \dfrac{p}{q}$,

$$D_0(x)^q = x^p + \alpha_2\ x^{2p}(-Ln\ x) + \cdots + \alpha_N\ x^{Np}\ (-Ln\ x) + \psi_k(x),$$

where ψ_k is $\mathcal{C}_1^k$, k-flat. Expanding $\psi \circ D_0 \circ \varphi$ and ordering the terms, we obtain that, for any k, there exists a sequence of coefficients $\lambda_i : \lambda_1 = r < \lambda_2 < \cdots < \lambda_{N(k)}$, $\lambda_{N(k)} \geq k$ and a sequence of polynomials $P_1 = A$ (a constant),$\ldots$, $P_{N(k)}$ such that

$$D(x) = \sum_{i=1}^{N(k)} x^{\lambda_i}\ P_i(Ln\ x) + \psi_k(x),$$

where ψ_k is a $\mathcal{C}^k$, k-flat function.

The coefficients λ_i, and the polynomials P_i are well defined, i.e., independent of k. This means that taking $k' > k$, the sequence for k is the sequence for k', truncated at order $N(k)$. This is similar to the unicity of the Taylor series. Taking k arbitrarily large, we have a well defined infinite series $\widehat{D}(x) = \displaystyle\sum_{i=1}^{\infty} x^{\lambda_i}\ P_i(x)$ which is asymptotic to $D(x)$ in the following sense:

For any $k \in \mathbb{N} - \{0\}$, $|\ D(x) - \sum_{i=1}^{s} x^{\lambda_i}\ P_i(Ln\ x)\ | = O(x^{\lambda_s})$ where $\lambda_1 = r < \lambda_2 < \cdots < \lambda_s < \cdots$ is an infinite sequence of positive coefficients tending to $+\infty$, and $P_1, P_2, \ldots, P_s, \ldots$ is an infinite sequence of polynomials.

The series $\widehat{D}$ is called the *Dulac series* of the map D.

A $\mathcal{C}^\infty$ function on $]0, X[$, which admits at $x = 0$ a series as above is said to be *quasi-regular*.

Remark 24 *For a Dulac series of the transition map, we have noticed that $\lambda_1 = r$, the hyperbolicity ratio, and $P_1 = 1$ so that $D(x) = Ax^r + O(x^r)$. It is also easy to verify that $\lambda_i \in \mathbb{N} + r\mathbb{N}$ for all i, and that P_i is constant for any i when $r \notin \mathbb{Q}$. Logarithmic terms occur only when the saddle is resonant.*

5.1.3.2 $D_\lambda(x)$ when $p = q = 1$

For the study in the next paragraph, we now want to write an expansion for $D_\lambda(x)$ in the case $p = q = 1$. We have to compute (5.40). To this end we need the following result:

Lemma 21 *Let $\varphi_\lambda(x)$ be a C^k parameter family of diffeomorphisms as above. Then, with the convention introduced in Section 1.2,*

$$\omega \circ \varphi_\lambda = c(\lambda)\omega + \cdots + \xi_k(x, \lambda), \tag{5.42}$$

where $c(\lambda) > 0$, for $\lambda \in W_1$ and ξ_k is C^k in (x, λ) and k-flat at $x = 0$ uniformly in $\lambda \in W_1$.

Proof. Let

$$\tilde{\omega} = \omega \circ \Phi_\lambda \quad , \quad \Phi_\lambda(x) = u(\lambda)x(1 + \cdots + \widetilde{\Phi}_k) \tag{5.43}$$

(again with the above convention) and $u(\lambda) > 0$.

$$\tilde{\omega} = \frac{u^{-\alpha_1} x^{-\alpha_1}(1 + \cdots \tilde{q}_k)^{-\alpha_1} - 1}{\alpha_1}$$

$$= u^{-\alpha_1} \frac{x^{-\alpha_1}(1 + \cdots + \widetilde{\Phi}_k)^{-\lambda_1} - 1}{\alpha_1} + \frac{u^{-\alpha_1} - 1}{\alpha_1},$$

$\varphi(\lambda) = \dfrac{u^{-\alpha_1} - 1}{\alpha_1}$ is a C^∞ function of λ,

$$\frac{x^{-\alpha_1}(1 + \cdots + \widetilde{\Phi}_k)^{-\alpha_1} - 1}{\alpha_1} = x^{-\alpha_1} \frac{(1 + \cdots + \widetilde{\Phi}_k)^{-\alpha_1} - 1}{\alpha_1} + \omega,$$

$\psi = \dfrac{(1 + \cdots \widetilde{\Phi}_k)^{-\alpha_1} - 1}{\alpha_1}$ is a C^k function in (x, λ) and $x^{-\alpha_1} = \alpha_1 \, \omega + 1$.

Finally, we obtain $\tilde{\omega} = u^{-\lambda_1}(1 + \psi(x, \lambda))\omega + \cdots$ which has the desired form, once expanded. $\qquad\qquad\square$

If we substitute $\Phi_\lambda(x)$ in $D_{\alpha(\lambda)}(x) = x + \alpha_1[x\omega + \cdots] + \cdots$ and use the above lemma, it is clear that we obtain a similar expansion to (5.34), but with new coefficients α_i which are now of class C^k.

Next, writing

$$\psi_\lambda(y) = \gamma_1(\lambda)y + \gamma_2(\lambda)y^2 + \cdots + \gamma_k(\lambda)y^k + O(y^{k+1}), \tag{5.44}$$

we obtain $\psi_\lambda \circ D_{\alpha(\lambda)} \circ \Phi_\lambda$ by substituting $D_{\alpha(\lambda)} \circ \Phi_\lambda$ into (5.44). It is clear that we can reorder the terms of this expansion to obtain a similar expansion to (5.34). We have proved

Proposition 12 *Let X_λ, σ, τ be as above and let $D_\lambda(x)$ be the transition map from σ to τ, with $r(\lambda_0) = 1$. Let W_1 be a neighborhood of λ_0 such that $D_\lambda(x)$ is defined from $\sigma \times W_1$ to τ. Then, there exists a sequence of neighborhoods of λ_0, $W_1 \supset W_2 \supset \cdots \supset W_k \supset \cdots$ such that for all $k \in \mathbb{N}$ there exist C^k functions $\alpha_1^k(\lambda), \ldots, \alpha_k^k(\lambda)$ and an expansion*

$$D_\lambda(x) = x + \sum_{i=1}^{k} \alpha_i^k(\lambda)[x^i \, \omega + \cdots] + \psi_k(x, \lambda) \tag{5.45}$$

for $(x, \lambda) \in W_k$, with the conventions as in Theorem 2. Here $\alpha_1^k \equiv \alpha_1 = r(\lambda) - 1$ for any k.

5.1.3.3 Mourtada's form for D_λ

The expression (5.45) will be used to study unfoldings of homoclinic saddle loop (see next section). To study hyperbolic polycycles with more than 1 singular point, A. Mourtada has introduced a simpler expression, which is valid without any assumption on $r(\lambda_0)$.

We consider transversal segments $\sigma = [0, X]$, $\tau = [-Y, Y]$ as above and let $D_\lambda(x) : \sigma \times W_0 \to \tau$ be the transition map.

Definition 25 *(1) Let $W_k \subset W_0$ be a neighborhood of λ_0 and I_k be the set of functions $f : [0, X] \times W_k \to \mathbb{R}$ with the following properties:*

(i) *f is C^∞ on $]0, X] \times W_k$.*

(ii) *$\varphi_j(x, \lambda) = x^j \dfrac{\partial^j f}{\partial x^j} (x, \lambda) \to 0$ for $x \to 0$, for each $j \le k$, uniformly on λ (we will say that $\dfrac{\partial^s f}{\partial x^j} = o(x^{-j})$ uniformly in λ).*

(2) A function $f : [0, X] \times W_0 \to \mathbb{R}$ is said to be of class I if f is C^∞ on $]0, X] \times W_0$ and if for each k there exists a neighborhood $W_k \subset W_0$, of $\lambda_0 \in P$ such that f is of class I_k on W_k.

Theorem 15 *(Mourtada [M1]). Let X_λ, σ, τ, D_λ be as above. Then, for $(x, \lambda) \in \sigma \times W_0$,*

$$D(x, \lambda) = x^{r(\lambda)} (A(\lambda) + \Phi(x, \lambda)), \qquad (5.46)$$

with $\Phi \in I$ and $A(\lambda)$ a C^∞ positive function.

Proof. We just explain briefly the ideas of the proof. The details can be found in [M1]. First, we note that, for given transversal segments σ and τ with their parametrizations, the functions A and Φ in formula (5.46) are *unique.* This follows from the fact that $r(\lambda)$ is well defined, that

$$A(\lambda) = \lim_{x \to 0} x^{-r(\lambda)} D(x, \lambda),$$

and also that $\Phi(x, \lambda) = x^{-r(\lambda)} D(x, \lambda) - A(\lambda)$.

Next, we note that a function is of the form (5.46), where $A(\lambda)$ is a C^k function and $\Phi \in I_k$ if and only if this is true after compositions to the right and to the left by C^k families of diffeomorphisms $\phi_\lambda(x)$ and $\psi_\lambda(x)$, with $\phi_\lambda(0) = \psi_\lambda(0) = 0$ for all $\lambda \in W_0$.

Therefore it suffices to prove that, in C^k-normal form coordinates

$$D_{\alpha(\lambda)}(x) = x^{r(\lambda)} (B(\lambda) + \psi(x, \lambda)), \qquad (5.47)$$

with $B(\lambda)$ a C^k function and $\psi \in I_k$.

To prove this, we consider two cases:

(i) if $r(\lambda_0)$ is irrational, then $D_{\alpha(\lambda)}(x) = x^{r(\lambda)}$ and the result is trivial,

(ii) if $r(\lambda_0) = \dfrac{p}{q}$, we apply Theorem 14, at some order $k' >> k$. In fact, we must note that the sum $+ \cdots$ in the first bracket begins by a monomial $x^{sp}\,\omega^{\ell}$ with $s \geq 2$, $\ell \leq p$. This is also the case for the other brackets.

Now, we can write for any k',

$$(D_{\alpha(\lambda)}(x))^q = x^p + \alpha_1(\lambda)x^{p\omega} + \sum_{\substack{\ell \,\leq\, s \,\leq\, K \\ s \geq 2}} \alpha_{s\ell}\,(\lambda)x^{sp}\,x^{\ell} + \psi_K(x,\lambda), \quad (5.48)$$

where the coefficients $\alpha_{s\ell}(\lambda)$ are $C^{k'}$, $K(k') \in \mathbb{N}$ and ψ_K is $C^{k'}$, k'-flat at $x = 0$.

Note that $x^p + \alpha_1(\lambda)x^{p\omega} = x^{qr(\lambda)}$. We can rewrite (5.48) in the following form:

$$D_{\alpha(\lambda)}(x) = x^{r(\lambda)}\left[1 + \sum_{\substack{\ell \,\leq\, s \,\leq\, K \\ s \geq 2}} \alpha_{s\ell}(\lambda)x^{sp-qr(\lambda)}\,\omega^{\ell}\right.$$

$$\left. + x^{-qr(\lambda)}\,\psi_K(x,\lambda)\right]^{\frac{1}{r(\lambda)}}. \quad (5.49)$$

Choosing k' large enough, $x^{-qr(\lambda)}\,\psi_K$ is C^k, k-flat at $x = 0$.

Now, $sp - qr(\lambda) \geq q\left(2\,\dfrac{p}{q} - r(\lambda)\right)$. For $\lambda = \lambda_0$, $q\left(2\,\dfrac{p}{q} - r(\lambda_0)\right) = 2p$.

If we take λ belonging to some small neighborhood W_k, of λ_0, then there exists $c > 0$ such that $q\left(2\,\dfrac{p}{q} - r(\lambda)\right) > c$.

Using the Taylor series of the function $u \to (1 + u)^{\frac{1}{r(\lambda)}}$ gives the desired form for $D_{\alpha(\lambda)}(x)$. $\hfill\square$

Definition 26 *We call expression (5.46), Mourtada's form of the transition map, and call $\mathcal{D}_k$ (respectively $\mathcal{D}$) the class of maps as in (5.46), when $\Phi \in I_k$ (resp. I).*

The importance of the classes I and $\mathcal{D}$ comes from the following theorem, which is easily proved by direct computation.

Theorem 16 *(Mourtada [M1]).*

(i) *I is an algebra.*

(ii) *If $f \in I$ and $g \in \mathcal{D}$, then $f \circ g \in I$.*

(iii) *Maps of class $\mathcal{D}$ can be composed. More precisely, if*

$$D_{1,\lambda}(x) = D_1(x,\lambda) = x^{r_1(\lambda)}\,(A_1(\lambda) + \Phi_1(x,\lambda))$$

and

$$D_{2,\lambda}(x) = D_2(x,\lambda) = x^{r_2(\lambda)}(A_2(\lambda) + \Phi_2(x,\lambda)),$$

then

$$D_{2,\lambda} \circ D_{1,\lambda}(x) = D_{3,\lambda}(x) = x^{r_3(\lambda)}(A_3(\lambda) + \Phi_3(x,\lambda)),$$

with

$$r_3(\lambda) = r_2(\lambda).r_1(\lambda), A_3(\lambda) = A_2(\lambda).A_1(\lambda)^{r_2(\lambda)}$$

and $\Phi_3 \in I$. *Hence,* $D_{3,\lambda} \in \mathcal{D}$.

(iv) *If* $D_\lambda \in \mathcal{D}$, *then* $D_\lambda^{-1} \in \mathcal{D}$.

(v) *I is closed under the derivation* $x \dfrac{d}{dx}$, *if* $f \in I$ *then* $x \dfrac{df}{dx} \in I$. *As a consequence, if* $D_\lambda = x^{r(\lambda)}(A(\lambda) + \Phi(x,\lambda)) \in \mathcal{D}$, *then*

$$\frac{\partial D_\lambda}{\partial x} = r(\lambda)x^{r(\lambda)-1}(A(\lambda) + \psi(x,\lambda)) \quad , \text{ with } \psi \in I. \tag{5.50}$$

(vi) *Any smooth germ is in I; any smooth diffeomorphism germ g at 0, with $g(0) = 0$ is in $\mathcal{D}$.*

5.2 Unfoldings of saddle connections in the finite codimension case

Let X_λ be a $\mathcal{C}^\infty$ family of vector fields in S such that X_{λ_0} has a saddle connection Γ. We want to study the unfolding defined by X_λ along $\Gamma \times \{\lambda_0\}$. Hence, we can suppose that X_λ is restricted to some neighborhood $\mathcal{U}$ of Γ in S, diffeomorphic to an annulus (S is supposed to be orientable) and λ belongs to some neighborhood W_0, of λ_0 in the parameter space. As above, let σ, τ be some transversal sections near the saddle point s of X_{λ_0}, we choose a local system of coordinates (x,y) in a neighborhood V of s, such that $s = (0,0)$ is the saddle point of X_λ for all $\lambda \in W_0$, and that $0x$, $0y$ are the unstable and stable local manifolds and σ, τ are parametrized respectively by $x \in [0, X[$ and $y \in]-Y, Y[$. We assume to have chosen a section $\sigma' \supset \bar{\sigma}$, parametrized by $]-X', X''[$ (see Figure 5.2).

The Poincaré map $P_\lambda(x) : \sigma \times W_0 \to \sigma' =]-X', X''[$ may be obtained as the composition

$$P_\lambda(x) = R_\lambda^{-1} \circ D_\lambda(x), \tag{5.51}$$

where R_λ is the $\mathcal{C}^\infty$ regular transition map from σ' to τ for $-X_\lambda$ (we assume W_0 and X chosen small enough such that P_λ is defined on $\sigma \times W_0$).

5.2.1 The codimension 1 case

As a generic assumption for codimension 1 bifurcations we can suppose that $r(\lambda_0) \neq 1$. Using the notation and the results of Section 1.3, $T_\lambda(\lambda) = R_\lambda^{-1} - \beta(\lambda)$, where $\beta(\lambda) = R_\lambda^{-1}(0)$, and D_λ are in $\mathcal{D}$. This gives

$$P_\lambda(x) = x^{r(\lambda)}(A(\lambda) + \Phi(x,\lambda)) + \beta(\lambda), \quad \text{with } \Phi \in I, \tag{5.52}$$

$r(\lambda_0) \neq 1$ and $A(\lambda_0) > 0$.

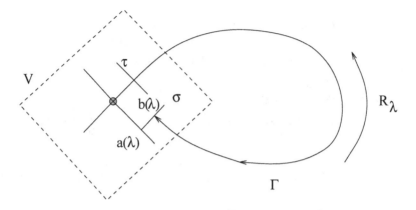

Figure 5.2

It follows from Theorem 16 that

$$\frac{\partial P_\lambda}{\partial x} = r(\lambda)x^{r(\lambda)-1}\left(A(\lambda) + \Phi_1(x,\lambda)\right).\qquad(5.53)$$

Therefore $\dfrac{\partial P_\lambda}{\partial x} \to 0$ (resp. ∞), for $x \to 0$ (resp. ∞), uniformly in $\lambda \in W_0$ for $r(\lambda_0) > 1$ (resp. $r(\lambda_0) < 1$). From Rolle's theorem, it now follows that the equation $\{P_\lambda(x) - x = 0\}$ has at most one root in σ, for any $\lambda \in W_0$, if σ is small enough.

Now, the positive roots x, sufficiently small, correspond to periodic orbits whose Hausdorff distance to Γ is sufficiently small. This is quite an obvious generalization of Lemma 6 in 4.1, that we now formulate without proof for general limit periodic sets.

Lemma 22 *Let Γ be any limit periodic set for a family X_λ at the parameter value λ_0. Let σ, σ' the transversal sections to Γ as above. Let $P_\lambda(x)$ be the Poincaré map of X_λ from σ to σ' (we suppose that λ belongs to some neighborhood W_0 of λ_0 in the parameter space).*

Let $\delta_\lambda(x) = P_\lambda(x) - x$. Then, for each $\varepsilon > 0$, we can find a neighborhood $\sigma(\varepsilon) \subset \sigma$ of $x_0 \in \Gamma \cap \sigma$ such that $x \in \sigma(\varepsilon)$ is a root of $\{\delta_\lambda(x) = 0\}$ for $\lambda \in W_0$ if and only if the orbit γ of X_λ through x is a periodic orbit with $d_H(\gamma, \Gamma) \le \varepsilon$ (d_H is the Hausdorff distance corresponding to a chosen distance in the phase space).

The computation of the cyclicity of Γ is equivalent to the computation of the number of roots of δ_λ, on $\sigma(\varepsilon)$ for ε and small enough W_0. The fact that $\delta_\lambda(x) = P_\lambda(x) - x$ has at most one root on $W_0 \times \sigma$ implies

Proposition 13 *Let* Γ *be a saddle connection as above, with* $r(\lambda_0) \neq 1$. *Then* $\mathcal{C}ycl(X_\lambda, \Gamma) \leq 1$.

In fact, we can deduce a more precise result from (5.52). It is always possible to construct a $\mathcal{C}^\infty$ 1-parameter family $\widetilde{X}_\beta$, near $\Gamma \times \{0\}$ with the return map

$$\widetilde{P}_\beta(x) = x^r + \beta, \tag{5.54}$$

where $(r-1)(r(\lambda_0) - 1) > 0$. For instance, we can take a fixed linear vector field in the coordinate chart with transition map $\widetilde{D}(x) = x^r$ and glue this chart with a second chart near a regular arc on $\Gamma - \{s\}$ so that the transition $\widetilde{R}^{-1}(y) = y + \beta$ (see [IY1]). Now, it is not difficult to prove that the two mappings $P_\lambda(x)$ and $\widetilde{P}_{p(\lambda)}(x)$ are $\mathcal{C}^\circ$-conjugate for all λ. Next, we can extend this conjugacy to an equivalence between $\widetilde{X}_{p(\lambda)}$ and X_λ, for each $\lambda \in W_0$ in some neighborhood of Γ. It is even possible (but more difficult) to obtain an equivalence depending continuously on λ [AAD]:

Theorem 17 *Let* $\widetilde{X}_\beta$ *be an unfolding of saddle connection with return map* $\widetilde{P}_\beta(x) = x^r + \beta$, *where* $(r-1)(r(\lambda_0)-1) > 0$. *Then, the unfolding* (X_λ, Γ) *is induced by the map* $\beta(\lambda)$, *up to* $(\mathcal{C}^\circ, \mathcal{C}^\circ)$-*equivalence* $(\widetilde{X}_\beta$ *is a versal unfolding of* Γ).

The bifurcation diagram of $\widetilde{X}_\beta$ is quite simple. We can suppose for instance that $r > 1$ (if not, change $\widetilde{X}_\beta$ by $-\widetilde{X}_\beta$), a hyperbolic stable limit cycle bifurcates from Γ for positive β (see Figure 5.3).

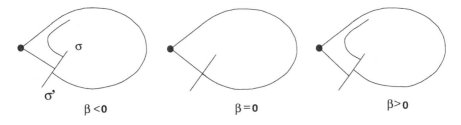

$\beta < 0$ $\qquad\qquad\qquad\qquad$ $\beta = 0$ $\qquad\qquad\qquad\qquad$ $\beta > 0$

Figure 5.3

Of course, if X_λ is already a 1-parameter family such that $\beta'(\lambda_0) \neq 0$, we can replace λ by β, up to a diffeomorphism in the parameter space, and the diagram of bifurcation of X_λ is the same as the one of $\widetilde{X}_\beta$.

Theorem 18 *Let* (X_λ, Γ) *be a 1-parameter unfolding of a saddle connection* Γ, *for the value* λ_0, *which verifies the generic assumptions* $r(\lambda_0) \neq 1$ *and* $\beta'(\lambda_0) \neq 0$.

Then, the unfolding (X_λ, Γ) *is* $(\mathcal{C}^\circ, \mathcal{C}^\circ)$-*equivalent to the "model"* $(\bar{X}_\beta, \Gamma)$. *In particular, this unfolding is unique, up to a* $(\mathcal{C}^\circ, \mathcal{C}^\circ)$ *equivalence and up to the change of* X_λ *by* $-X_\lambda$. *It is structurally stable.*

5.2.2 The k-codimension case, $k \geq 2$

From now on we suppose that $r(\lambda_0) = 1$. A saddle connection with this condition is of course of codimension greater than 2 (one condition is needed to express the connection and another one is $r(\lambda_0) = 1$). To make precise the notion of codimension, we will used the so-called Dulac expansion for the return map $P(x) = P_{\lambda_0}(x)$ for X_{λ_0} along Γ.

Using Proposition 12 for D_λ and the Taylor expansion of $R_\lambda(x)$, putting $\delta_\lambda(x) = P_\lambda(x) - x$, we have

$$\begin{aligned}
\delta_\lambda(x) = \beta_0(\lambda) \quad &+ \quad \alpha_1(\lambda)[x\omega + \cdots] \\
&+ \quad \beta_1(\lambda)x + \alpha_2(\lambda)[x^2\omega + \cdots] + \cdots \\
&+ \quad \beta_{k-1}(\lambda)x^{k-1} + \alpha_k(\lambda)[x^k\omega + \cdots] + \cdots \psi_k(x,\lambda),
\end{aligned} \tag{5.55}$$

for any order $k \in \mathbb{N}$. Here, $\beta_0(\lambda) = P_\lambda(0) = b(\lambda) - a(\lambda)$, where $a(\lambda)$, $b(\lambda)$ are the first intersections of the unstable and the stable manifold of s with σ'. We have that $\alpha_1(\lambda) = 1 - r(\lambda)$, where $r(\lambda)$ is the hyperbolicity ratio and $\alpha_1(0) = \beta_0(0) = 0$.

The functions β_i come from the Taylor expansion of $R_\lambda(x)$ and are $\mathcal{C}^\infty$, but the $\alpha_i(\lambda)$ come from the formula for D_λ. They depend in general on k and are just $\mathcal{C}^k$.

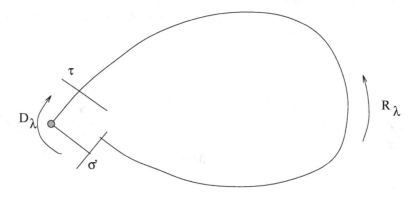

Figure 5.4

Taking $\lambda = \lambda_0$ in the formula (5.55), we obtain the expansion

$$\delta_{\lambda_0}(x) = \beta_1 x + \alpha_2 x^2(-Lnx) + \cdots$$

$$+\beta_{k-1}\, x^{k-1} + \alpha_k\, x^k(-Lnx) + 0(x^k). \tag{5.56}$$

Note that the coefficients β_i, α_j we obtain in this way are independent of k; if we write a similar expansion at order $k' \geq k$ for $\delta_{\lambda_0}(x)$, then the coefficients α_i,

β_j for $i \leq k$, $j \leq k - 1$ are the same. So, using expansion (5.56) at any order k, we obtain a well defined series

$$\widehat{\delta}_{\lambda_0}(x) = \sum_{i=1}^{\infty} \left(\beta_i \, x^i + \alpha_{i+1} \, x^{i+1} \, (-Lnx) \right), \qquad (5.57)$$

which is asymptotic to $\delta_{\lambda_0}(x)$ in the following sense:

$$\left| \delta_{\lambda_0}(x) - \sum_{i=1}^{k} \left(\beta_i x^i + \alpha_{i+1} \, x^{i+1}(-Lnx) \right) \right| = o(x^k), \qquad (5.58)$$

for any $k \in \mathbb{N} - \{0\}$.

We call this series, the Dulac series for $\delta_{\lambda_0}(x)$.

Definition 27 *Let $k \geq 1$. We say that the saddle connection Γ of X_{λ_0} is of codimension $2k$ if $\delta_{\lambda_0}(x) \sim \beta_k \, x^k$ with $\beta_k \neq 0$, and of codimension $2k + 1$ if $\delta_{\lambda_0}(x) \sim \alpha_{k+1} \, x^{k+1} Lnx$ with $\alpha_{k+1} \neq 0$.*

With this definition we have

Theorem 19
Let (X_λ, Γ) be a $\mathcal{C}^\infty$ unfolding of codimension $\ell \geq 2$. Then $\mathcal{C}ycl(X_\lambda, \Gamma) \leq \ell$.

To study the number of zeros of δ_λ, we have to extend somewhat the algebra generated by the monomials $x^i \omega^j$. We now introduce the algebra of functions continuous in (x, λ) which are finite combinations of the monomials $x^{\ell + n\alpha_1} \omega^n$, with $\ell, n \in \mathbb{Z}$, $m \in \mathbb{N}$, $\alpha_1 = \alpha_1(\lambda)$. The coefficients are continuous functions of λ. We call it the *algebra of admissible functions.*

Of course, we consider also these monomials as functions of (x, α_1), but when we consider combinations of monomials, α_1 is always replaced by the function $\alpha_1(\lambda)$.

Now, we introduce between the monomials, the following *partial* strict order:

$$x^{\ell' + n'\alpha_1} \, \omega^{m'} \prec x^{\ell + n\alpha_1} \, \omega^m \iff \begin{cases} \ell' < \ell \text{ or} \\ \ell' = \ell, \ n' = n \text{ and } m' > m. \end{cases}$$

Notice that $x^{\ell + n'\lambda_1} \, \omega^{m'}$ and $x^{\ell + n\alpha_1} \, \omega^m$ with $n \neq n'$ are not ordered.

Later on, the notation $f + \cdots$ where f is a monomial will mean that after the sign $+$ there is a (unwritten) finite combination of monomials g_i with $g_i \succ f$. (This definition extends the one used in Theorem 2.)

We will also use the symbol $*$ to replace any continuous function of λ, non-zero at $\lambda = \lambda_0$, and we write $\dot{\Phi}$ for the derivative with respect to x : $\dot{\Phi} = \dfrac{\partial \Phi}{\partial x}$. With these conventions, we now indicate some easy properties of the algebra of admissible functions.

a) Let g, f be two monomials with $g \succ f$, then $\frac{g}{f}(x, \alpha_1) \to 0$, for $(x, \alpha_1) \to$
$(0, 0)$. This follows from the two following observations: $\omega \geq \operatorname{Inf}\left(\frac{1}{|\alpha_1|}, -Lnx\right)$
and $x^{s(\alpha_1)} \omega^m \to 0$ (for any continuous function $s(\alpha_1)$, with $s(0) > 0$), if $(x, \alpha_1) \to$
$(0, 0)$ and $m \in \mathbb{N}$.

b) Let $f \succ 1$ be a monomial, then $f(x, \alpha_1) \to 0$, for $x \to 0$ (uniformly in α_1);
$f \succ 1$ means that $f = x^{\ell + n\alpha_1} \omega^m$ with $\ell \geq 1$, and we can use the same argument
as in a).

c) Let $f_1 \succ f_2$. Then $gf_1 \succ gf_2$, for any g.

d) Let $f = x^{\ell + n\alpha_1} \omega^m$, then

$$\dot{f} = (\ell + (n - m)\alpha_1)x^{\ell - 1 + n\alpha_1} \omega^m - mx^{\ell - 1 + n\alpha_1} \omega^{m-1}.$$

From this formula it follows easily that

e) Let be $f = x^{\ell + n\alpha_1} \omega^m$ with $\ell \neq 0$ and g any monomial such that $g \succ f$.
Then $\dot{g}$ is a combination of two monomials g' and g'' and $\dot{f} = * f' + \cdots$ with
$g' \succ f', g'' \succ f'$.

We shall also use rational functions of the algebra of the following type:
$\frac{f + \cdots}{1 + \cdots}$. We call them *admissible rational functions*.

For them, we have

f) $\left(\dfrac{x^{\ell + n\alpha_1} \omega^m}{1 + \cdots\cdots\cdots}\right)^{\bullet} = * \dfrac{x^{\ell - 1 + n\alpha_1} \omega^m}{1 + \cdots\cdots\cdots}$ if $\ell \neq 0$.

We can now give a proof of Theorem 19. We consider successively the two
cases α_{k+1} and β_k.

Proof of Theorem 19 in the case α_{k+1}.

Recall that we can write

$$\delta_\lambda(x) = \beta_0 + \alpha_1[x\omega + \cdots] + \beta_1 x + \alpha_2[x^2\omega + \cdots] + \cdots$$

$$+\alpha_k[x^k \omega + \cdots] + \beta_k x^k + \alpha_{k+1} x^{k+1} \omega + \cdots + \psi_K, \tag{5.59}$$

where α_i, β_j are continuous functions, ψ_K is a C^K function of (x, α), K-flat in x,
with $K > 2k + 1$. We suppose that

$$\beta_0(\lambda_0) = \cdots = \beta_k(\lambda_0) = 0, \quad \alpha_1(\lambda_0) = \cdots = \alpha_k(\lambda_0) = 0 \quad \text{and} \quad \alpha_{k+1}(\lambda_0) \neq 0.$$

From the property d) above it follows that

$$(x^j \omega)^{\bullet} = (j - \alpha_1)x^{j-1} \omega + \cdots \quad \text{if } j \neq 0 \quad \text{and} \quad \dot{\omega} = x^{-1-\alpha_1}.$$

Differentiating δ_λ, we obtain, also using property e),

$$\dot{\delta}_\lambda = \alpha_1 [*\omega + \cdots] + \beta_1 + \alpha_2 [*x\omega + \cdots] + \cdots + *\alpha_{k+1} \, x^k \, \omega + \cdots + \dot{\psi}_K. \quad (5.60)$$

If we differentiate δ_λ, $k+1$ times, we find

$$\delta_\lambda^{(k+1)} = \alpha_1 [* \, x^{-k-\alpha_1} + \cdots] + \alpha_2 [* \, x^{-k+1-\alpha_1} + \cdots] + \cdots$$

$$+ * \, \alpha_{k+1} \, \omega + \cdots + \psi_K^{(k+1)}. \quad (5.61)$$

All the monomials $\beta_j x^j$, for $j \leq k$ have disappeared. Multiplying by $x^{k+\alpha_1}$, we obtain (using property c)):

$$x^{k+\alpha_1} \, \delta_\lambda^{(k+1)} = [*1 + \cdots] + \alpha_2 [*x + \cdots] + \cdots$$

$$+ * \, \alpha_{k+1} \, x^{k+\alpha_1} \, \omega + \cdots + x^{k+\alpha_1} \, \psi_K^{(k+1)}. \quad (5.62)$$

Above and in what follows each bracket designates an admissible function.

Locally, in some neighborhood of $(\lambda_0, 0)$, the zeros of $\delta_\lambda^{(k+1)}$ are zeros of the following function $\xi_1 = \dfrac{x^{k+\alpha_1} \, \Delta_\lambda^{(k+1)}}{[* \, 1 + \cdots \cdots]}$, where the denominator is the first bracket in (5.62):

$$\xi_1 = \alpha_1 + \alpha_2 \frac{* \, x + \cdots}{* \, 1 + \cdots} + \cdots$$

$$+ \alpha_k \, \frac{* \, x^{k-1} + \cdots}{* \, 1 + \cdots} + \frac{* \, \alpha_{k+1} \, x^{k+\alpha_1} \, \omega + \cdots}{* \, 1 + \cdots \cdots} + \Phi_1. \quad (5.63)$$

Here $\Phi_1 = \dfrac{x^{k+\alpha_1} \, \psi_K^{(k+1)}}{* \, 1 + \cdots \cdots}$ is a C^{K-k-1} function, at least $(K - k - 1)$-flat at $x = 0$. Using property f), we have

$$\dot{\xi}_1 = \alpha_2 \, \frac{* \, 1 + \cdots}{* \, 1 + \cdots} + \cdots$$

$$+ \alpha_k \, \frac{* \, x^{k-2} + \cdots}{* \, 1 + \cdots \cdots} + \frac{* \, \alpha_{k+1} \, x^{k-1+\alpha_1} \, \omega + \cdots}{* \, 1 + \cdots \cdots} + \Phi_2, \quad (5.64)$$

where $\Phi_2 = \dot{\Phi}_1$ is C^{K-k-2}, $(K - k - 2)$-flat in $x = 0$; $\dot{\xi}_1 = \alpha_2 \, u_1 + \cdots$, where u_1 is invertible as a rational admissible function. Let $\xi_2 = u_1^{-1} \, \dot{\xi}_1$. Differentiating ξ_2 again we obtain

$$\dot{\xi}_2 = \alpha_3 \, \frac{* \, 1 + \cdots}{* \, 1 + \cdots} + \cdots + \dot{\Phi}_2. \quad (5.65)$$

We have $\dot{\xi}_2 = \alpha_3 \, u_2 + \cdots$, where u_2 is invertible as an admissible rational function. We define $\xi_3 = u_2^{-1} \, \dot{\xi}_2$, and so on.

In this way, we find a sequence of functions $\xi_1, \ldots, \xi_k$ such that ξ_j is the product of $\dot{\xi}_{j-1}$ by some invertible admissible rational function. For the last ξ_k, we have

$$\xi_k = \alpha_k + \frac{* \ \alpha_{k+1} \ x^{1+\alpha_1} \ \omega + \cdots}{* \ 1 + \cdots \cdots} + \Phi_k, \tag{5.66}$$

where Φ_k is C^{K-2k}, $(K-2k)$-flat.

Differentiating one last time, we obtain

$$\dot{\xi}_k = \frac{* \ \alpha_{k+1} \ x^{\alpha_1} + \cdots}{* \ 1 + \cdots \cdots} + \dot{\Phi}_k. \tag{5.67}$$

Then, using the fact that $\dot{\Phi}_k$ is $(K-2k-1)$-flat, with $K-2k-1 > 0$, and the property a), we obtain

$$x^{-\alpha_1} \ \omega^{-1} \dot{\xi}_k = +\alpha_{k+1} + o(1). \tag{5.68}$$

Here $o(1)$ denotes a function $\psi(x, \lambda)$, such that $\psi(x, \lambda) \to 0$, for $x \to 0$, uniformly in λ. The hypothesis $\alpha_{k+1} \ (\lambda_0) \neq 0$ implies that locally $x^{-\alpha_1} \omega^{-1} \dot{\xi}_k$ and also $\dot{\xi}_k$ are non-zero for (λ, x), $x \geq 0$. Therefore the function ξ_k has at most one zero, for (x, λ) near $(0, \lambda_0)$; ξ_{k-1} has at most two zeros, and so on. ξ_1 has at most k zeros, locally. Now, ξ_1 has at least the same number of zeros as $\delta_\lambda^{(k+1)}$, and finally the function $\delta_\lambda(x)$ has at most $2k+1$ zeros near 0, for λ near λ_0.

Proof in the case β_k.

We differentiate the map δ_λ only k times,

$$\delta_\lambda^{(k)}(x) = \alpha_1 [* \ x^{-k+1-\alpha_1} + \cdots] + \cdots$$

$$+\alpha_k [* \ \omega + \cdots] + *\beta_k + \cdots + \psi_K^{(k)}. \tag{5.69}$$

Next, we introduce

$$\xi_1 = \frac{\delta_\lambda^{(k)}(x)}{[* \ x^{-k+1-\alpha_1} + \cdots]} = \alpha_1 + \alpha_2 \ \frac{* \ x + \cdots}{* \ 1 + \cdots} + \cdots$$

$$+\frac{* \ \alpha_k \ x^{k-1+\alpha_1} \ \omega + * \ \beta_k \ x^{k-1+\alpha_1} + \cdots}{* \ 1 + \cdots \cdots} + \Phi_1, \tag{5.70}$$

where Φ_1 is C^{K-k}, $(K-k)$-flat at $x = 0$.

As in the previous case, we define a sequence of functions $\xi_1, \ldots, \xi_{k-1}$ with ξ_j equal to $\dot{\xi}_{j-1}$ multiplied by an invertible admissible rational function. The last function ξ_{k-1} is equal to

$$\xi_{k-1} = * \ \alpha_{k-1} + \frac{* \ \alpha_k \ x^{k-1+\alpha_1} \ \omega + * \ \beta_k \ x^{1+\alpha_1} + \cdots}{* \ 1 + \cdots \cdots} + \Phi_{k-1}, \tag{5.71}$$

so

$$\dot{\xi}_{k-1} = \frac{* \, \alpha_k \, x^{\alpha_1} \, \omega + * \, \beta_k \, x^{\alpha_1} + \cdots}{* \, 1 + \cdots \cdots} + \dot{\Phi}_{k-1}, \tag{5.72}$$

where $\dot{\Phi}_{k-1}$ is C^{K-2k+1} , $(K - 2k + 1)$-flat.

We now define a function ξ_k,

$$\xi_k = x^{-\alpha_1} \, \omega^{-1} [* \, 1 + \cdots] \dot{\xi}_{k-1} = * \, \alpha_k + * \, \beta_k \, \frac{* \, 1 + \cdots}{* \, 1 + \cdots} \cdot \frac{1}{\omega} + \Phi_k, \tag{5.73}$$

where the bracket is the denominator in (5.72). The function Φ_k is C^{K-2k}, $(K - 2k)$-flat.

Differentiating ξ_k, we obtain

$$\dot{\xi}_k = * \, \beta_k \, \frac{x^{-1-\alpha_1} + \cdots}{* \, 1 + \cdots \cdots} \cdot \frac{1}{\omega^2} + \dot{\Phi}_k, \tag{5.74}$$

and

$$\omega^2 \, \frac{* \, 1 + \cdots}{* \, x^{-1-\lambda_1} + \cdots} \, \dot{\xi}_k = * \, \beta_k + \omega^2 \, \frac{* \, 1 + \cdots}{* \, x^{-1-\lambda_1} + \cdots} \cdot \dot{\Phi}_k. \tag{5.75}$$

The remainder is $o(1)$. Hence $\dot{\xi}_k \neq 0$, for (λ, x) sufficiently near $(\lambda_0, 0)$, as $\beta_k(\lambda_0) \neq 0$. It follows, as in the previous case, that the map Δ_λ has at most $2k$ zeros for (λ, x) near $(\lambda_0, 0)$.

5.2.3 Bifurcation diagrams for generic unfoldings

To cover the two cases more easily, we call now β_i, all the coefficients in the δ_λ-expansion. So we write

$$\delta_\lambda(x) = \beta_0(\lambda) + \beta_1(\lambda)[x\omega + \cdots] + \cdots$$

$$+\beta_{2m-1}(\lambda)[x^m \, \omega + \cdots] + \beta_{2m} \, (\lambda)x^m + \cdots + \psi_k. \tag{5.76}$$

Now, even coefficients correspond to the monomials x^m and odd ones to monomials $[x^m \, \omega + \cdots]$.

Suppose that X_{λ_0} has a saddle connection Γ of finite codimension $n > 0$, $\beta_0(\lambda_0) = \cdots = \beta_{n-1} \, (\lambda_0) = 0$ and $\beta_n(\lambda_0) \neq 0$.

Then, the following result was proved by P. Joyal [J2]:

Theorem 20 *There exists a neighborhood* $[0, \varepsilon] \times W_0$ *of* $(0, \lambda_0)$ *and a mapping* $\alpha(\lambda) = (\alpha_0(\lambda), \dots, \alpha_{n-1} \, (\lambda))$ *defined on* W_0 *such that the zeros of the polynomial*

$$P(x, \alpha) = \alpha_0 + \alpha_1 x + \cdots + \alpha_{n-1} \, x^{n-1} + x^n$$

in $[0, \varepsilon]$ *are those of (5.76).*

Remark 25 *In fact, Joyal looked at* $\Delta_\lambda(x) = D_\lambda(x) - R_\lambda(x)$ *in place of* $\delta_\lambda(x) = P_\lambda(x) - x$, *but the two equations* $\Delta_\lambda = 0$ *and* $\delta_\lambda = 0$ *are equivalent.*

If X_λ is a generic family with ℓ parameters $(\dim P = \ell)$, we know that any saddle connection Γ at λ_0 has codimension $n \leq \ell$ and also that the map $\beta(\lambda) = (\beta_0(\lambda) \cdots \beta_{n-1}(\lambda))$ is of maximal rank at λ_0. In the case that $\ell = n$, Joyal proved

Theorem 21 *If* $\beta(\lambda) = (\beta_0(\lambda), \ldots, \beta_{n-1}(\lambda))$ *is a mapping of maximal rank (i.e.,* $\dfrac{D(\beta_0, \ldots, \beta_{n-1})}{D(\lambda_1, \ldots, \lambda_n)}(\lambda_0) \neq 0$), *then there exists a neighborhood* $[0, \varepsilon] \times W_0$ *of* $(0, \lambda_0)$, *and a homeomorphism* $\alpha(\lambda)$ *such that the zeros of (5.76) in* $[0, \varepsilon]$ *are the same as the zeros of the polynomial* $P(x, \alpha)$ *in* $[0, \varepsilon]$.

Remark 26

1) *The proof of Theorem 21 given by Joyal is rather difficult to understand. It would be important to find a clearer one.*

2) *It is easy to construct a generic unfolding of codimension* n *by gluing charts (as explained in the codimension 1 case). Theorem 20 says that such an unfolding is versal for any other unfolding of codimension* n *saddle connection.*

3) *It is reasonable to think that Theorem 21 extends to any generic unfolding; if* $n \leq \ell$, *we expect the existence of a topological submersion* $\alpha(\lambda)$ *near* λ_0 *of the parameter space onto the parameter space* $(\alpha_0, \ldots, \alpha_{n-1})$. *Then the local bifurcation diagram should be equal to a topological trivial product of the diagram with* n *parameters.*

The proof given by Joyal for these two theorems are based on the notion of the Tchebychef family.

Definition 28 *Let* $f_0, \ldots, f_n$ *be real functions on some interval* $[a, b] \subset \mathbb{R}$. *One says that* $\{f_0, \ldots, f_n\}$ *is a Tchebychef system if any combination* $\displaystyle\sum_{i=0}^{n} \alpha_i \, f_i, \alpha_i \in \mathbb{R}$ *has no more than* n *isolated zeros on* $[a, b]$. *It is called* regular *if any subsystem* $f_0, \ldots, f_k$, *for* $k < n$, *is also Tchebychef.*

Remark 27 *The number of zeros is counted with multiplicity. We want to consider also functions like* $x^k Ln^\ell x$, *non-differentiable at the ends* a, b *of the interval. Joyal introduced a general definition of multiplicity for such non-differentiable functions.*

For a function with a Dulac series expansion $\displaystyle\sum_{\ell \leq k} a_{ij} x^k Ln^\ell x$, *the multiplic-ity at 0 is defined using the well ordered sequence of the monomials* $x^k Ln^\ell x$: *we say that the multiplicity at zero is* n *if the first non-zero coefficient in the Dulac expansion is of order* n.

Now, the following well-known result for differentiable systems was extended by Joyal for the general system [J1].

Theorem 22 *Let $\{f_0, \ldots, f_n\}$ and $\{g_0, \ldots, g_n\}$ be regular systems in $[a, b]$. Let $\mu = (\mu_0, \ldots, \mu_{n-1})$ and $\nu = (\nu_0, \ldots, \nu_{n-1})$. If $\mu_n \neq 0$ and $\nu_n \neq 0$, then there exists a homeomorphism $\mu(\nu)$ such that $\mu_0 \ f_0(x) + \cdots + \mu_{n-1} \ f_{n-1}(x) + \mu_n \ f_n(x)$ and $\nu_0 \ g_0(x) + \cdots + \nu_{n-1} \ g_{n-1}(x) + \nu_n \ g_n(x)$ have the same zeros in $[a, b]$.*

In particular, $\{1, x, \ldots, x^n\}$ is a regular system in any interval $[a, b]$. The theorem implies that the bifurcation diagram of the zeros of $\sum\limits_{i=0}^{n} \mu_i f_i$, $\mu_n \neq 0$, for any regular Tchebychef system, is homeomorphic to the diagram of the polynomial family $x^n + \sum\limits_{i=0}^{n-1} a_i x^i$.

Of course, we can apply these results for the germs at one end of the interval.

Any strictly increasing sequence of n monomials $x^i \ Ln^j \ x$, with $0 \leq j \leq i$, and ordered by increasing flatness, is a regular Tchebychef system for any n. It may be proved by a "differentiation-division" argument as in the proof of Theorem 19.

To prove Theorem 20, we can consider the extended unfolding

$$\Delta_{\beta, \lambda, u} = \beta_0 \ f_0 + \beta_1 \ f_1 + \cdots + \beta_{n-1} \ f_{n-1} + f_n, \tag{5.77}$$

where $f_i(x, u, \lambda)$ is the corresponding function factor of β_i where we put $\omega(x, u) = \dfrac{x^{-u} - 1}{u}$ and consider f_i as an independent function of $x, u, \lambda, \beta_0, \ldots, \beta_n$, and $f_n(x, u, \lambda) = \beta_n[\cdots] + \psi_k$ includes the last term in the expansion of Δ and the remaining term ψ_k.

It is also a consequence of Theorem 19 that the sequence $\{f_0, \ldots, f_n\}$ is a regular Tchebychef system, and so the bifurcation diagram Σ of $\delta_{\beta, \lambda, u}$ is topologically transversal to the β-planes: $(u, \lambda) = \mathcal{C}$onstant.

The map δ_λ is equal to

$$\delta_\lambda = \delta_{\beta(\lambda), u(\lambda)}. \tag{5.78}$$

The bifurcation diagram of Δ_λ is just the counter-image $\varphi^{-1}(\Sigma)$, where $\varphi(\lambda) = (\beta(\lambda), \lambda), u(\lambda))$.

To prove Theorem 21, we have to prove that the map φ is (topologically) transversal to Σ. This is not the case for *any* map φ. For instance, look at

$$\delta_{u, \beta}(x) = \beta_0 + \beta_1 \ x\omega + x.$$

It is easy to prove that for $u = 0$,

$$\delta_{0, \beta}(x) = \beta_0 + \beta_1 x(-Lnx) + x$$

has a bifurcation diagram on $[0, \infty]$ with a line d of double zeros and a sector S with 2 simple zeros (see Figure 5.5).

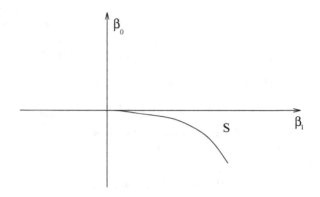

Figure 5.5

If we take $\lambda = (\beta_0, \beta_1)$ and define the map $\varphi(\lambda) = (\beta_0, \beta_1, u = \beta_1)$, we obtain

$$\delta_\lambda(x) = \beta_0 + x^{1-\beta_1},$$

which has just 1 zero on $[0, \infty]$ for each $\lambda \in \mathbb{R}^2$.

The proof of this point, i.e., the transversality of φ to the bifurcation diagram Σ, is a very obscure point in the proofs of Joyal's theorems, and it is really difficult to check it. I just want to give the direct and independent proof for $n = 2$ which appeared in [DRS1].

So let the following be a generic expansion of δ_λ at order 2:

$$\delta_\lambda(x) = \beta_0(\lambda) + \beta_1(\lambda)x\omega + \beta_2(\lambda)x + \gamma_1(\lambda)x^2\omega^2 + \gamma_2(\lambda)x^2\omega + \psi(x, \lambda), \quad (5.79)$$

with $\lambda \in \mathbb{R}^2 \to (\beta_0(\lambda), \beta_1(\lambda))$ a local diffeomorphism at λ_0 and $0 < \beta_2(\lambda_0) < 1$. Locally, we can assume that $\lambda = (\beta_0, \beta_1)$ in a neighborhood of 0 and $\omega = \dfrac{x^{-\beta_1} - 1}{\beta_1}$. Put $\xi_\lambda = \beta_1\omega + \cdots$ Equation $\delta_\lambda(x) = 0$ can be written

$$-\beta_0 = \xi_\lambda(x), \quad (5.80)$$

where $\xi_\lambda(x)$ is such that

$$\dot{\xi}_\lambda(x) = \beta_1[(1 - \beta_1)\omega - 1] + \beta_2 + o(1). \quad (5.81)$$

To study the bifurcation diagram we consider different cases.

(i) Case $\beta_1 \geq 0$. Here $x^{-\beta_1} - 1 \geq 0$ and, because of our hypothesis, $\beta_2(0) > 0$ and $\beta_1(0) = 0$. We see that

$$\exists A > 0, \ \exists U > 0, \quad \text{such that}$$

$$\forall(\beta_1, x) \in [0, A] \times [0, U] : \dot{\xi}_\lambda > 0.$$

Indeed, for $\beta_1 > 0 : \dot{\xi}_\lambda \to \infty$ when $x \to 0$; for $\beta_1 = 0: \dot{\xi}_\lambda \to \beta_2$ when $x \to 0$ (see Figure 5.6).

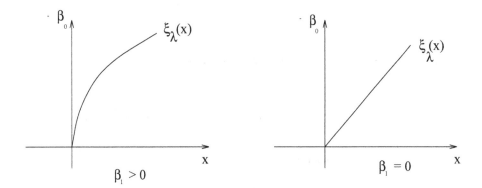

Figure 5.6

For $\beta_0 \geq 0$, no $x > 0$ is a solution of $\xi_\lambda(x) = -\beta_0$.
For $\beta_0 < 0$, there exists a unique solution $x > 0$ of $\xi_\lambda(x) = -\beta_0$.
The bifurcation diagram for $\beta_1 \geq 0$ is given in Figure 5.7.

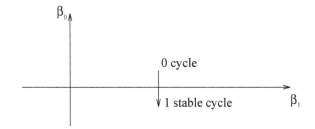

Figure 5.7

For each $\beta_1 \geq 0$ fixed we have the creation of one stable limit cycle when β_0 goes from positive to negative values; $\{\beta_0 = 0, \beta_1 \geq 0\}$ is a half line of saddle connections with non-zero divergence at the saddle point (codimension 1 bifurcation).

(ii) Case $\beta_1 < 0$.

$$(x^2\,\omega^2)^\bullet = (2-2\beta_1)x\,\omega^2 - 2x\,\omega,$$
$$(x^2\,\omega)^\bullet = (2-\beta_1)x\,\omega - x.$$

Hence,

$$\dot{\xi}_\lambda = (1 + \beta_1)(x^{-\beta_1} - 1) + \beta_2 - \beta_1 \quad + \quad \gamma_1[(2 - 2\beta_1)x\ \omega^2 - 2x\ \omega]$$
$$+ \quad \gamma_2[(2 - \beta_1)x\ \omega - x] + \dot{\psi}.$$

As $(x\ \omega^2)^\bullet = (1 - 2\beta_1)\omega^2 - 2\omega$,

$$\ddot{\xi}_\lambda = -\beta_1(1 - \beta_1)x^{-\beta_1 - 1} + \gamma_1[(2 - 2\beta_1)[(1 - 2\beta_2)\omega^2 - 2\omega] - 2(1 - \beta_1)\omega + 2]$$

$$\Big[+\gamma_2[(2 - \beta_1)(1 - \beta_1)\omega - (2 - \beta_1) - 1] + \ddot{\psi},$$

$$\ddot{\xi}_\lambda = -\beta_1(1 - \beta_1)x^{-\beta_1 - 1} \quad + 2\gamma_1(1 - \beta_1)(1 - 2\beta_1)\omega^2 + (1 - \beta_1)[-6\gamma_1 + (2 - \beta_1)\gamma_2]\omega$$

$$+ [2\gamma_1 + (\beta_1 - 3)\gamma_2] + \ddot{\psi}.$$

We can find a bounded function $O(1)$ such that

$$\ddot{\xi}_\lambda = -\beta_1(1 - \beta_1)x^{-1 - \beta_1} + \omega^2\ O(1).$$

From this, we will prove that $\exists A' > 0$, $\exists U' > 0$, such that for all $\beta_1 \in]-A', 0[$, the graph of ξ_λ on $[0, U']$ looks like that in Figure 5.8.

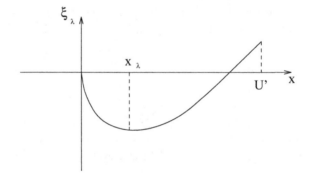

Figure 5.8

We see that x_λ is a unique strict minimum, $\dot{\xi}_\lambda(x_\lambda) = 0$, $\ddot{\xi}_\lambda(x_\lambda) > 0$, $\dot{\xi}_\lambda > 0$ on $]x_\lambda, U'[$ and $\dot{\xi}_\lambda < 0$ on $[0, x_\lambda[$ with $\dot{\xi}_\lambda \to -\infty$ for $x \to 0$.

Therefore, consider

$$\frac{1}{\omega^2}\ \ddot{\xi}_\lambda = -\frac{\beta_1(1 - \beta_1)}{\omega^2\ x^{1+\beta_1}} + O(1).$$

$$\omega = \frac{x^{-\beta_1} - 1}{\beta_1} \leq |\ Lnx\ | \text{ for } U' < 1 \text{ and } A' \text{ sufficiently small.}$$

Choosing some $\delta > 0$, taking $-\dfrac{\delta}{2} \leq \beta_1 < 0$, and U' sufficiently small,

$$\frac{1}{\omega^2} \ddot{\xi}_\lambda \geq \frac{|\beta_1|}{(Ln\ x)^2\ x^{1-\delta/2}} + O(1) \geq \frac{|\beta_1|}{x^{1-\delta}} + O(1).$$

Take $M > 0$ so that $O(1) \geq -M$ on $[0, U'] \times [-\delta, 0]$, then

$$\frac{1}{\omega^2} \ddot{\xi}_\lambda \geq M \text{ if } \frac{|\beta_1|}{x^{1-\delta}} \geq 2M \Longleftrightarrow x \leq \left(\frac{|\beta_1|}{2M}\right)^{\frac{1}{1-\delta}}.$$

Let us write $x \leq C\ |\ \beta_1\ |^\mu$ with $C = \frac{1}{(2M)^\mu}$ and $\mu = \frac{1}{1-\delta}$.

For these values of U', δ, M, let us consider $\dot{\xi}_\lambda$ on $[C\ |\ \beta_1\ |^\mu, U']$,

$$\dot{\xi}_\lambda = \beta_1(1 - \beta_1)\omega + \beta_2 - \beta_1 + o(1).$$

As $\beta_1 < 0$, the function $\beta_1 \omega = x^{-\beta_1} - 1 = e^{-\beta_1 Ln\ x} - 1$ is strictly increasing and negative, so that for all $x \in [C\ |\ \beta_1\ |^\mu, U']$,

$$\beta_1 \omega(C\ |\ \beta_1\ |^\mu) \leq \beta_1 \omega(x) \leq 0,$$

and

$$\beta_1 \omega(C\ |\ \beta_1\ |^\mu) = O(|\ \beta_1 Ln(C\ |\ \beta_1\ |^\mu)\ |)$$

$$\Rightarrow \beta_1 \omega(x) = O(|\ \beta_1 Ln(C\ |\ \beta_1\ |)Ln(C\ |\ \beta_1\ |^\mu)\ |) \text{ for } x \in [C\ |\ \beta_1\ |^\mu, U'].$$

Hence,

$$\dot{\xi}_\lambda = (1 - \beta_1)O(|\ \beta_1\ |\ Ln(C\ |\ \beta_1\ |^\mu)) + \beta_2 - \beta_1 + o(1).$$

For U' sufficiently small, $|\ o(1)\ | \leq \beta_{2/3}$, and for A' sufficiently small,

$$|-\beta_1 + (1 - \beta_1)O(|\ \beta_1 Ln(C\ |\ \beta\ |^\mu)\ |)\ | \leq \beta_{2/3},$$

implying that $\dot{\xi}_\lambda \geq \beta_{2/3} > 0$.

As a conclusion, we see that for $-A' \leq \beta_1 < 0$ fixed, there is a bifurcation value (corresponding to a generic coalescence of limit cycles) for $\xi_\lambda(x_\lambda) = -\beta_0(\lambda)$. This bifurcation occurs at $\beta_0 = \Gamma(\beta_1)$, Γ is a $\mathcal{C}^\infty$ function, for $\beta_1 < 0$, because of the implicit function theorem applied to the Poincaré mapping of the $\mathcal{C}^\infty$ vector field X_λ in the neighborhood of the semi-stable limit cycle (see Figure 5.9).

Taking $\Gamma(0) = 0$, we will now say something about the behavior of $\Gamma(\beta_1)$ in the neighborhood of $\beta_1 = 0$:

$$x_\lambda \text{ is given by } \dot{\xi}_\lambda(x_\lambda) = 0$$

$$\text{i.e., } \beta_1(1 - \beta_1)\omega(x_\lambda) + \beta_2 - \beta_1 + o(1) = 0,$$

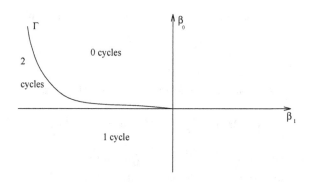

Figure 5.9

or equivalently,

$$| \beta_1 | \, \omega(x_\lambda) = \frac{1}{1 - \beta_1} \, (\beta_2 - \beta_1 + o(1)).$$

Now,

$$\beta_0 = x_\lambda \Big[| \beta_1 | \, \omega(x_\lambda) - (\beta_2 + o(1)) \Big]$$

$$= x_\lambda \Big[\frac{\beta_1 \, \beta_2}{1 - \beta_1} - \frac{\beta_1}{1 - \beta_1} + o(1) \Big] \le x_\lambda.$$

And, as $| \beta_1 | \, | Ln \, x_\lambda | \ge | \beta_1 | \, \omega(x_\lambda) \ge \frac{\beta_2}{2}$ for A' and U' sufficiently small we see that

$$x_\lambda \le e^{-\frac{\beta_2}{2|\beta_1|}},$$

and hence

$$\Gamma(\beta_1) \le e^{-\frac{\beta_2}{2|\beta_1|}}. \tag{5.82}$$

That is Γ is ∞-flat, at $\beta_1 = 0$.

Remark 28

1) Above we just proved that $\Gamma(\beta_1)$ is ∞-flat at 0 in a C°-sense (formula (5.82)). In fact, Γ is C^∞ at 0 and is ∞-flat in a C^∞-sense (all derivatives tend to zero when $\beta_1 \to 0$)[R7].

2) Unfoldings of codimension k saddle connections appear in cusp unfoldings of codimension $k + 1$ (the Bogdanov-Takens bifurcation corresponds to $k = 1$). In [R2] it is proved that these unfoldings have finite cyclicity. Next, P. Joyal [J3] and P. Mardesic [Mar3] proved that generic unfoldings of finite codimension cusp singularities are versal and gave a complete description of their bifurcation diagrams. The saddle connection unfoldings appear inside the cusp unfoldings as perturbations of Hamiltonian vector fields. In this context an independent proof of theorems similar to Theorems 20, 21 above was given by P. Mardesic in [Mar1], [Mar3].

5.3 Unfoldings of saddle connections of infinite codimension

In this section, we restrict ourselves to analytic unfoldings (X_λ, Γ), where Γ is a homoclinic saddle connection for the parameter value λ_0. As in the smooth case, we can suppose that the same point s is a saddle point, for any λ belonging to some neighborhood W_0 of λ_0, and that there exists an analytic chart (x, y), where $s = (0,0)$ and the axis $0x$, $0y$ are respectively the unstable and stable local manifolds, for X_λ, $\lambda \in W_0$. Let σ', τ be analytic transversal segments to $0y$, $0x$, σ' parametrized by $x \in] - X', X'[$, with $0 \leq X < X'$, τ by $y \in] - Y, Y[$; let be $\sigma = [0, X[$.

Take X, Y small enough such that $D_\lambda : \sigma \times W_0 \to \tau$ and $R_\lambda^{-1} : \tau \times W_0 \to \sigma'$ are defined. Let $P_\lambda(x) = R_\lambda^{-1} \circ D_\lambda : \sigma \times W_0 \to \sigma'$ be the Poincaré return map and $\delta_\lambda(x) = P_\lambda(x) - x : \sigma \times W_0 \to \mathbb{R}$ be the displacement function.

Definition 29 *We will say that the unfolding (X_λ, Γ) is of infinite codimension if the Dulac series $\hat{\delta}_{\lambda_0}$ is equal to zero.*

For smooth families this condition means that δ_{λ_0} is C^∞ and flat at $x = 0$. For analytic families we have seen in Chapter 3 that $\delta_{\lambda_0}(x)$ is quasi-analytic and the condition $\hat{\delta}_{\lambda_0} \equiv 0$ is equivalent to $\delta_{\lambda_0} \equiv 0$. This means that the nearby trajectories, on the side where the return map is defined, are periodic orbits; the vector field X_{λ_0} is of center type.

5.3.1 Finite cyclicity property of analytic unfoldings

Analytic unfoldings of homoclinic connections are studied in [R5]. Here we just want to explain the principal steps of this study and complete it with some new results (Theorem 25 below). We refer the reader to [R5] for more details.

Firstly, we need a version of the asymptotic expansion at order k, given in formula (5.45) for analytic unfoldings.

Proposition 14 *Let σ, σ' be analytic sections as above and let X_λ be an analytic unfolding with $r(\lambda_0) = 1$ (we do not assume that $\delta_{\lambda_0}(x) \equiv 0$). Then, for any $k \in \mathbb{N} - \{0\}$, there exist neighborhoods of λ_0, $W_k \subset W_0$ and analytic maps $\beta_0^k, \ldots, \beta_k^k$, $\alpha_1^k, \ldots, \alpha_{k+1}^k$ from W_k to $\mathbb{R}$, such that*

$$\delta_\lambda(x) = \beta_0^k + \alpha_1^k[x\omega + \cdots] + \cdots + \beta_k^k x^k + \alpha_{k+1}^k x^{k+1} \omega + \psi_k(x, \lambda) \qquad (5.83)$$

on $\sigma \times W_k$, where ψ_k is C^k, k-flat at $x = 0$. Expressions in brackets are finite combinations of monomials $x^i \omega^j$, $0 \leq j \leq i \leq k$, with coefficients analytic in λ, zero at λ_0. Any monomial in $+ \cdots$ has an order strictly larger than the order of the leading term.

We cannot use the proof given in the smooth case because the change of coordinates we use to reduce X_λ in normal form has only finite differentiability. As a consequence the coefficients β_i^k also have only finite differentiability.

We replace Theorem 13 used for studying smooth families with the following result, from Dulac:

Theorem 23 *(Dulac normal form) For any $N \in \mathbb{N}$, up to an analytic equivalence, X_λ is equal to*

$$X_\lambda = x\,\frac{\partial}{\partial x} \; - y\Big(1 + \sum_{j=1}^{N} \alpha_i(\lambda)(xy)^i + (xy)^N\, G(x, y_i, \lambda)\Big)\,\frac{\partial}{\partial y}, \qquad (5.84)$$

where α_i, G are analytic functions.

Based on this Dulac normal form, we can establish Proposition 14 (see [R5]).

In fact, we have (and need!) a stronger result: the Dulac normal form is valid for holomorphic families and we can apply it to some complex extension of X_λ (in some neighborhood $\widehat{V}$ of Γ in $\mathbb{C}^2$) and for some neighborhood $\widehat{W}_0$ of λ_0 in $\mathbb{C}^\Lambda$ (if the real parameter space is $\mathbb{R}^\Lambda$). Using this complex extension $\widehat{X}_{\widehat{\lambda}}$ of X_λ we can prove that $\delta_\lambda(x)$ also has a complex extension: σ' may be extended into a disk $\widehat{\sigma}'$ in $\mathbb{C}$ and σ into a sector $\widehat{\sigma}$ at $0 \in \mathbb{C}$, containing σ (see Figure 5.10).

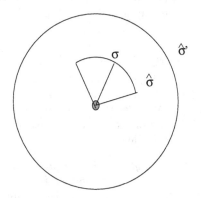

Figure 5.10

Then, there exists a map $\widehat{\delta}_{\widehat{\lambda}}(\widehat{x}) : \widehat{\sigma} \times \widehat{W}_0 \to \mathbb{C}$ with a similar expansion as (5.83), where $\widehat{x} \in \widehat{\sigma}$, $\widehat{\omega}$ is a holomorphic extension of ω on $\widehat{\sigma} \times \widehat{W}_0$ and the coefficients $\widehat{\beta}_i^k$, $\widehat{\alpha}_j^k$ are holomorphic on $\widehat{W}_0$:

$$\widehat{\delta}_{\widehat{\lambda}}(\widehat{x}) = \widehat{\beta}_{\widehat{\theta}}^k + \widehat{\alpha}_1^k\,[\widehat{x}\,\widehat{\omega} + \cdots] + \cdots + \widehat{\psi}_k\,(\widehat{x}, \widehat{\lambda}). \qquad (5.85)$$

The remainder $\widehat{\psi}_k : \widehat{\sigma} \times \widehat{W}_0 \to \mathbb{C}$ is just C^k-real and k-flat at $\widehat{x} = 0$ (but of course holomorphic like all the other terms, on $\widehat{\sigma} - \{0\}$).

From now on, we want to take into account that (X_λ, Γ) is an ∞-codimension unfolding. In the previous chapter we introduced the Bautin Ideal $\mathcal{I}$. This ideal is generated by the coefficient germs $\widetilde{a}_i$ of the series expansion $\delta_\lambda(x) = \sum_i a_i(\lambda)(x - x_0)^i$ at any *regular* point $x_0 \in \sigma - \{0\}$.

Our interest in the holomorphic extension comes from the fact that it allows us to pass to the limit $x_0 = 0$.

Lemma 23 *Assume an expansion of $\delta_\lambda(x)$ of order k, as in Proposition 7, is given. Then each coefficient germ $\widetilde{\beta}_i^k, \widetilde{a}_j^k$ belongs to $\mathcal{I}$.*

Proof. First, differentiating expansion (5.85) $k + 1$ times, we eliminate all the monomials x^i, $i \leq k$,

$$\widehat{\delta}_{\widehat{\lambda}}^{(k+1)}(\widehat{x}) = \widehat{a}_1^k [x^{-k-\alpha_1} + \cdots] + \cdots + + \widehat{a}_{k+1}^k \, \widehat{\omega} + \widehat{\psi}_k^{(k+1)}. \tag{5.86}$$

Divide this expression by the first bracket $x^{-k-\alpha_1}[*1 + \cdots]$:

$$\frac{x^{k+\alpha_1} \, \widehat{\delta}_{\widehat{\lambda}}^{(k+1)}(\widehat{x})}{[*1 + \cdots\cdots]} = \widehat{a}_1^k + \widehat{a}_2^k \, \frac{[* \, x + \cdots]}{[*1 + \cdots]} + \cdots = \widehat{a}_1^k + o(1). \tag{5.87}$$

The term $o(1) \to 0$ for $\widehat{x} \to 0$ (uniformly in $\widehat{\lambda}$).

For each $\widehat{x} \in \widehat{\sigma} - \{0\}$, the germ $\widehat{\lambda} \to \widehat{\delta}_{\widehat{\lambda}}(\widehat{x})$ at λ_0 belongs to the ideal $\widehat{\mathcal{I}}$ obtained by the complexification of the Bautin Ideal $\mathcal{I}$. But, as this complex ideal $\widehat{\mathcal{I}}$ is *closed*, the germ of $(\widehat{\delta}_{\widehat{\lambda}}^{(k+1)}(x), \lambda_0)$ also belongs to $\widehat{\mathcal{I}}$ and this is also true for the left member of (5.87) and its limit for $\widehat{x} \to 0$. This means that $(\widehat{a}_1^k, \lambda_0) \in \widehat{\mathcal{I}}$.

We can prove in the same way that any germ $(\widehat{a}_i^k, \lambda_0) \in \widehat{\mathcal{I}}$.

For instance, if $\widehat{\delta}_1$ is the left member of (5.87), we have that $(\widehat{\delta}_1 - \widehat{a}_1^k, \lambda_0)$ and also $(\widehat{a}_2^k(\lambda), \lambda_0) \in \widehat{\mathcal{I}}$, as

$$\frac{[*1 + \cdots]}{[* \, x + \cdots]} (\widehat{\delta}_1 - \widehat{a}_1^k) = \widehat{a}_2^k + o(1). \tag{5.88}$$

Next, we subtract the "α"-part of the expansion (5.86):

$$\widehat{\delta}_{\widehat{\lambda}}(\widehat{x}) - \sum_{i=1}^k \widehat{a}_i^k(\widehat{\lambda})[* \, \widehat{x}^{i+1} \, \widehat{\omega} + \cdots] = \sum_{j=0}^k \widehat{\beta}_j^k(\lambda)\widehat{x}^j + o(x^k). \tag{5.89}$$

The first part of the proof implies that the left hand side of (5.89) has a germ at λ_0 which belongs to $\widehat{\mathcal{I}}$ for any $\widehat{x} \neq 0$. Using that $\widehat{\mathcal{I}}$ is closed we prove as above that $(\widehat{\beta}_j^k, \lambda_0) \in \widehat{\mathcal{I}}$, for each $j \leq k$. $\qquad\square$

The preceding lemma means that the principal part of (5.85) belongs to $\widehat{\mathcal{I}}$, for any $\hat{x} \neq 0$. As $\widehat{\delta_\lambda}$ also belongs to $\widehat{\mathcal{I}}$, it follows that the remainder term $\widehat{\psi}_k(\hat{x}, \hat{\lambda})$ belongs to $\widehat{\mathcal{I}}$, too. This permits us to divide the expansion (5.85) in the ideal $\widehat{\mathcal{I}}$.

Let $\varphi_1, \ldots, \varphi_\ell$ be a system of generators of $\mathcal{I}$ and $\widehat{\varphi}_1, \ldots, \widehat{\varphi}_\ell$ their complex extensions which form a system of generators of $\widehat{\mathcal{I}}$. We can divide $\widehat{\delta_\lambda}$ (of order k),

$$\widehat{\delta}_\lambda(\hat{x}) = \sum_{i=1}^{\ell} \widehat{\varphi}_i \, \hat{h}_i^k \tag{5.90}$$

by

$$\hat{h}_i^k = \widehat{\beta}_{i0}^k + \hat{\alpha}_{i1}^k \left[\hat{x}\hat{\omega} + \cdots\right] + \cdots + \hat{\alpha}_{i,k+1}^k \, \hat{x}^{k+1} \, \hat{\omega} + \widehat{\psi}_i^k (\hat{x}, \hat{\lambda}) \tag{5.91}$$

where $\widehat{\beta}_{ij}^k$, $\hat{\alpha}_{ij}^k$ are holomorphic functions of $\hat{\lambda}$ and $\widehat{\psi}_i^k (\hat{x}, \hat{\lambda})$ is $\mathcal{C}^k$, k-flat at $\hat{x} = 0$.

We establish formula (5.90) using the theorem (D), as in the regular case in Chapter 4 (see the proof of Proposition 3 in 4.2). Finally, restricting (5.90) to δ_λ real we get

Theorem 24 *Let $\varphi_1, \ldots, \varphi_\ell$ be analytic functions on W_0 whose germs $\widetilde{\varphi}_1, \ldots \widetilde{\varphi}_\ell$ generate the Bautin Ideal $\mathcal{I}$. Let $k \in \mathbb{N}$, $k \geq 1$. There exists a neighborhood W_k of λ_0 (in W_0) and functions $h_i^k(x, \lambda)$, $1 \leq i \leq \ell$, with (ω, x)-expansion of order k,*

$$h_i^k(x, \lambda) = \beta_{i0}^k + \alpha_{i1}^k(\lambda)[x\omega + \cdots] + \cdots + \alpha_{i,k+1}^k \, x^{k+1} \, \omega + \psi_i^k(x, \lambda), \tag{5.92}$$

as in Proposition 14. These functions are factors of the division of $\delta(x, \lambda)$ in $\varphi_1, \ldots, \varphi_\ell$. This means that on $[0, x_0] \times W_k$ for some $x_0 > 0$, we have

$$\delta(x, \lambda) = \sum_{i=1}^{\ell} \varphi_i(\lambda) h_i^k(x, \lambda). \tag{5.93}$$

As in Chapter 4, we now restrict ourselves to some *minimal set of generators* for $\mathcal{I}$. Recall that the factors $h_i^k(x, \lambda_0)$ are independent of the choice of k. We thus have functions $h_i(x)$, defined on σ by $h_i(x) = h_i^k(x, \lambda_0)$, where $h_i^k(x, \lambda)$ are factor functions defined in Theorem 14. Taking the value at λ_0 for the expansions of $h_i^k(x, \lambda_0)$ for arbitrarily large k, we see that at $x = 0$ $h_i(x)$ has a well-defined Dulac series:

$$D^\infty h_i(x) = \sum_{j=0}^{\infty} \beta_{ij} \, x^j + \alpha_{ij+1} \, x^{j+1} \, Lnx. \tag{5.94}$$

$D^\infty \, h_i(x)$ is the series of $h_i(x)$ in the following sense:

$$\left| h_i(x) - \left(\sum_{j=0}^{N} \beta_{ij} \, x^i + \alpha_{ij+1} \, x^{j+1} \, Lnx\right) \right| = o(x^N) \tag{5.95}$$

for $\forall N \in \mathbb{N}$.

The functions h_i are analytic on $\sigma - \{0\}$ and are $\mathbb{R}$-independent. This implies that each of them is non-identically zero.

We can conjecture that they are quasi-analytic. If this were the case, then $h_i(x) \not\equiv 0 \Rightarrow D^\infty h_i(x) \not\equiv 0$, but we do not know if this is true. What was proved in [R5] was a weaker result which was sufficient to obtain the finite cyclicity. Here, we want to make this result precise by giving an explicit bound. To express it, we consider quasi-regular functions: they are functions f, analytic on $\sigma - \{0\}$ which at 0 has a Dulac series more general than (5.94):

$$D^\infty f = \sum_{0 \leq \ell \leq s} \alpha_{s\ell} \, x^s \, Ln^\ell \, x. \tag{5.96}$$

The monomials $x^s Ln^\ell x$ are totally ordered: $1 \prec xLnx \prec x \prec x^2 Ln^2 x \prec x^2 Lnx \prec x^2 \prec \cdots$. We call the *order* of a monomial $x^s Ln^\ell x$ its order in the above sequence: order $(x^s Ln^\ell x) = \frac{1}{2}s(s+3) - \ell$; for instance order $(1) = 0, \ldots$, order $(x^s) = \frac{1}{2}s(s+3)$.

We say that the order at 0 of a quasi-regular function is k if the first non-zero coefficient in $D^\infty f$ is of order $k+1$ in the list (the constant term is of order 0). We write $\text{order}_0(f) = k$. Of course, we say that $\text{order}_0(f) = \infty$ if $D^\infty f \equiv 0$ (this is equivalent to saying that f is ∞-flat at 0). We write $D^N f$ for the Dulac series, truncated at order $N \in \mathbb{N}$ (the sum of the $N+1$ terms).

Using this order for quasi-regular functions, it is possible to propose a definition of an index $s_\delta(0)$ which extends the index $s_\delta(u_0)$ defined for regular points in Chapter 4.

Let a minimal set of generators be given and let $h_1(x), \ldots, h_\ell(x)$ be the corresponding factors at $\lambda = \lambda_0$. Let s be the dimension of the vector space generated by the Dulac series $\hat{h}_i : 0 \leq s \leq \ell$. Notice that the series $D^\infty h_i = \hat{h}_i$ are $\mathbb{R}$-dependent $\Longleftrightarrow s < \ell$.

There exists some N such that $\dim\{D^N h_i\}_i = \dim\{D^\infty h_i\}_i$.

Definition 30

$$s_\delta(0) = \text{Inf } \{N \mid \dim\{D^N h_i\}_i = s\}.$$

The dimension s and the index $s_\delta(0)$ are independent of the choice of the minimal set of generators.

If $s = 0$, this means that any factor is flat. In this case we assume $s_\delta(0) = 0$, but we will prove that this cannot occur.

As we proved for a regular point in Lemma 4.9, it is possible to choose a minimal set of generators for $\mathcal{I}$ such that the corresponding non-flat factors h_i are in strictly increasing order,

$$\text{order }_0(h_1) < \cdots < \text{order}_0(h_s) < \infty$$

and order $(h_j) = \infty$ if $j \geq s + 1$.

We will say that this minimal set is *adapted to* 0. For such a minimal set, $s_\delta(0) = \text{order}_0(h_s)$. As a consequence, $s_\delta(0) \geq s - 1$.

The main result we want to prove in this section, which clarifies the finiteness result of [R5], is

Theorem 25 *Let X_λ be an analytic unfolding of a saddle connection Γ. Let σ be a section transverse to Γ, $\sigma \simeq [0, X]$ with $\{0\} = \Gamma \cap \sigma$.*

Let s, and $s_\delta(0)$ be as above. Then,

(i) $\mathcal{C}ycl(X_\lambda), \Gamma \leq s_\delta(0)$.

(ii) *If $\mathcal{I}$ is regular, $\mathcal{C}ycl(X_\lambda, \Gamma) \geq s - 1$.*

As in [R5] we introduced a desingularization map $\Phi : \widetilde{W}_0 \to W_0$ given by Hironaka's theory [H]. This map is a proper analytic map with the property that at each $\widetilde{\lambda}_0 \in D = \Phi^{-1}(\lambda_0)$, the lifted functions $\widetilde{\varphi}_i(\widetilde{\lambda}) = \varphi_i \circ \Phi(\widetilde{\lambda})$ have a monomial form,

$$\widetilde{\varphi}_i(\widetilde{\lambda}) = u_i(\widetilde{\lambda}) \prod_{i=1}^{\Lambda} z_1^{p_1^i} \cdots z_\Lambda^{p_\Lambda^i} \tag{5.97}$$

for local coordinates $(z_1, \ldots, z_n)$ at $\widetilde{\lambda}_0$; $\widetilde{u}_i(\widetilde{\lambda}_0) \neq 0$.

The germs of the $\widetilde{\varphi}_i$ at $\widetilde{\lambda}_0$ do not in general form a minimal set of generators of the lifted ideal $\widetilde{\mathcal{I}}^{\widetilde{\lambda}_0} = (\mathcal{I} \circ \Phi)^{\widetilde{\lambda}_0}$ at $\widetilde{\lambda}_0$, but it is possible to find a subsequence $i_1 < \cdots < i_L$ such that $\widetilde{\varphi}_{i_1}, \ldots, \widetilde{\varphi}_{i_L}$, have this property. Moreover, if $H_1, \ldots, H_L$ are the corresponding factors, then $\text{order}_0(H_j) = \text{order}_0(h_{i_j})$, $j = 1, \ldots, L$.

It follows that the corresponding index $s_{\widetilde{\delta}}^{\widetilde{\lambda}_0}$ at $0 \in \sigma$ for the lifted unfolding $X_{\Phi(\widetilde{\lambda})}$ at $\widetilde{\lambda}_0$ is less than $s_\delta(0)$. (We write $\widetilde{\delta}_{\widetilde{\lambda}}(x) = \delta_{\Phi(\widetilde{\lambda})}(x)$. This map is defined for $\widetilde{\lambda}$ in a neighborhood of $\widetilde{\lambda}_0$.)

By the definition of cyclicity, there exist a neighborhood $W(\widetilde{\lambda}_0)$ of $\widetilde{\lambda}_0 \in D$ and a value $x_{\widetilde{\lambda}_0}$ such that

$$\mathcal{C}ycl\ (X_{\Phi(\widetilde{\lambda})}, \ \Gamma \ ; \ \widetilde{\lambda}_0) = N(\widetilde{\lambda}_0),$$

where $N(\widetilde{\lambda}_0)$ is the maximal number of zeros for the equation $\{\widetilde{\delta}_{\widetilde{\lambda}}(x) = 0\}$ in $[0, x_{\widetilde{\lambda}_0}]$, for $\widetilde{\lambda} \in \widetilde{W}(\widetilde{\lambda}_0)$.

Extracting a finite subcovering from the covering $\{W(\widetilde{\lambda}_0)\}_{\widetilde{\lambda}_0}$ of the compact set D, we obtain that there exists some $\widetilde{\lambda}_0$ such that

$$\mathcal{C}ycl\ (X_\lambda , \ \Gamma \ ; \ \lambda_0) = N(\widetilde{\lambda}_0) = \mathcal{C}ycl(X_{\phi(\widetilde{\lambda})}, \ \Gamma \ ; \ \widetilde{\lambda}_0). \tag{5.98}$$

It suffices to choose $\widetilde{\lambda}_0$ such that

$$\mathcal{C}ycl\ (X_{\Phi(\widetilde{\lambda})}, \Gamma \ ; \widetilde{\lambda}_0) = \text{Sup}_{\widetilde{\lambda}_1 \in D}\ \mathcal{C}ycl\ (X_{\Phi(\widetilde{\lambda})}, \ \Gamma \ ; \ \widetilde{\lambda}_1).$$

Now, as we have noticed above, $s_{\widetilde{\delta}}^{\widetilde{\lambda_0}}(0) \le s_\delta(0)$. Hence, in order to prove part (i) of Theorem 13, it is sufficient to prove it at any $\widetilde{\lambda_0} \in D$.

Proof of part (i) of Theorem 25.

From now on, in this part of the proof, we drop the subscript. We just denote the family $X_{\Phi(\widetilde{\lambda})}$ by X_λ, $\widetilde{\lambda_0}$ by λ_0 and so on.

We assume to have chosen a minimal system of generators $\varphi_1, \ldots, \varphi_\ell$ (ℓ in place of L) of the special form (5.97). A consequence of this special form, proved in [R5], is the following:

Proposition 15 *If a minimal set of generators $\varphi_1, \ldots, \varphi_\ell$ has the form (5.97) and is adapted at 0, then there exists $0 < r \le 1$ such that*

$$W_0 = \bigcup_{i=1}^{s} V_i^r, \tag{5.99}$$

where

$$V_i^r = \{\lambda \in W_0 \mid \mid \varphi_i(\lambda) \mid \ge r \mid \varphi_j(\lambda) \mid, \ \forall j = 1, \ldots, \ell\}.$$

Remark 29 *This proposition means that the sets V_i^r, (which are related to the non-trivial series $\hat{h}_i(i \le s)$) are sufficient to cover a neighbourhood of λ_0.*

Sketch of proof of Proposition 15 (see details in [R5]).

Write $\varphi_i(\lambda) = u_i(\lambda)\, \psi_i(\lambda)$ where $u(\lambda_0) \ne 0$ and

$$\psi_i(\lambda) = \prod_{i=1}^{\Lambda} z_1^{p_1^i} \cdots z_\Lambda^{p_\Lambda^i} \ (\lambda = (z_1, \ldots, z_\Lambda) \ and \ \lambda_0 = (0, \ldots, 0)).$$

Let $W_i = \{\lambda \in W_0 \mid \mid \psi_i(\lambda) \mid \ge \mid \psi_j(\lambda) \mid \text{ for } \forall j \ne i\}$ and let also $I = \{i \mid \lambda_0 \in \text{int } \overline{W_i}\}$.

It is easy to prove that

(a) $U_{i \in I} W_i$ is a neighbourhood of λ_0 and also that $\overline{\text{int } W_i} = \overline{\overset{\circ}{W}_i}$, where $\overset{\circ}{W}_i = \{\lambda \in W_0 \mid \mid \psi_i(\lambda) \mid > \mid \psi_j(\lambda) \mid \text{ for } \forall j \ne i\}$. Next, if $i \in I$, it is possible to find an analytic arc $\lambda(\varepsilon) = (\varepsilon^{n_1}, \ldots, \varepsilon^{n_\Lambda})$, for some $n_1, \ldots, n_\Lambda$ such that

$$\text{order } (\psi_i \circ \lambda)_{|\varepsilon=0} < \text{order } (\psi_j \circ \lambda)_{|\varepsilon=0} \text{ for all } j \ne i.$$

Taking the division of $\delta(x, \lambda)$ given in Theorem 24, we have

$$\text{(b) } \delta(x, \lambda) = \sum_{i=1}^{\ell} \varphi_i(\lambda) h_i\,(x, \lambda).$$

Substituting $\lambda(\varepsilon)$ we obtain that

$$\text{(c)}\ \delta(x, \lambda(\varepsilon)) = \alpha_i\, h_i(x)\varepsilon^{n_i} + 0(\varepsilon^{n_i}),$$

for some $\alpha_i \neq 0$.

At this point we use a generalization of Il'yashenko's theorem (Theorem 3.7) saying that for any analytic unfolding of hyperbolic graphics (see below), every partial derivative of $\delta(x, \lambda)$ with respect to the parameter λ at $\lambda = \lambda_0$ is *also quasi-analytic* (for the proof see [MMR] and also [R5] for the 1-parameter case). The proof is quite similar to the one given in Chapter 3. We can apply it to the 1-parameter family $X_{\lambda(\varepsilon)}$ with displacement function $\Delta(x, \varepsilon) = \delta(x, \lambda(\varepsilon))$. We find that $h_i(x) = \dfrac{1}{\alpha_i}\dfrac{\partial^{n_i}}{\partial \varepsilon^{n_i}}\,\Delta(x, \varepsilon)_{\varepsilon=0}$ is quasi-analytic. As a consequence, we have that $I \subset \{1, \ldots, s\}$ (for each index in I, the corresponding function h_i is quasi-analytic, non-zero, and so $D^\infty h_i \not\equiv 0$). It follows from (a) that $\bigcup_{i=1}^{s} W_i$ is a neighborhood of λ_0 and that there exists a value r, $0 < r \leq 1$ such that $\bigcup_{i=1}^{s} V_i^r$ is also a neighborhood of λ_0. $\square$

We return now to the proof of Theorem 25. Let $f_1, \ldots, f_n, \ldots$ be the sequence of monomials $x^i\, \omega^j$, $0 \leq j \leq i$, indexed by order: $f_1 = 1$, $f_2 = x\omega$, and so on.

Let us be given a minimal set of generators, adapted to 0, and let $h_1, \ldots, h_\ell$ be the corresponding functions on σ. Let $n_i =$ order $h_i(0)$: $n_1 < n_2 \cdots < n_s < \infty$ and $n_j = \infty$, for $j \geq s+1$.

We will use the decomposition of the displacement function $\delta_\lambda(x)$ in the ideal generated by $\varphi_1, \ldots, \varphi_\ell$ at some high order of differentiability k, $(k >> s_\delta(0)$, the index at 0),

$$\delta_\lambda(x) = \sum_{i=1}^{\ell} \varphi_i(\lambda)h_i(x, \lambda) \tag{5.100}$$

(we do not write the superscript k).

For $N = s_\delta(0)$, we can write

$$h_i(x, \lambda) = \sum_{j=1}^{N} \alpha_{ij}\,(\lambda)f_j\,(x, \lambda) + \cdots + \psi_i\,(x, \lambda), \ i = 1, \ldots, \ell, \tag{5.101}$$

where $+\cdots$ is as usual a finite combination of monomials $x^i\omega^j$ and $\psi_i(x, \lambda)$ a C^k, k-flat function at $x = 0$.

We know that $\alpha_{ij}(\lambda_0) = 0$ if $j < n_i$, for $\forall i = 1, \ldots, \ell$.

We first want to rearrange the combination (5.100) in a new combination,

$$\delta_\lambda(x) = \sum_{i=1}^{N} \Phi_i(\lambda)\,H_i(x, \lambda), \tag{5.102}$$

with the following properties:

(i) $H_i(x, \lambda) = *f_i(x, \lambda) + \cdots + \tilde{\psi}_i(x, \lambda)$, $i = 1, \ldots, N$, and $\tilde{\psi}_i(x, \lambda)$, a C^k and k-flat function at $x = 0$.

(ii) If $W_i^r = \{\lambda \in W_0 \mid |\Phi_i(\lambda)| \geq r \mid \Phi_j(\lambda)|, \forall j = 1, \ldots, N\}$, then $\displaystyle\bigcup_{i=1}^{s} W_{n_i}^r = W_0$

in some new neighborhood $[0, x_0] \times W_0$.

We proceed in two steps:

(a) For any j, we modify φ_j in the combination

$$\varphi_j(\lambda) + \sum_{n_\ell < j} \beta_{j\ell}(\lambda)\varphi_{n_\ell}(\lambda) \ , \text{ where } \beta_{j\ell}(\lambda_0) = 0,$$

in such a way that in the new minimal system (5.101) holds with the extra condition $\alpha_{jn_p}(\lambda) \equiv 0$, for all $1 \leq p \leq s$ and $j > n_p$.

Clearly, the new minimal system $\varphi_1, \ldots, \varphi_\ell$ is adapted and if necessary taking new W_0 and r, we have preserved condition (5.99).

(b) We now define the new set of generators $\Phi_1, \ldots, \Phi_N$ (no longer a minimal one), by

$$\begin{aligned}
\Phi_1 &= \sum_{i=1}^{\ell} \varphi_i \, \alpha_{i1}, \ldots, \Phi_{n_1-1} = \sum_{i=1}^{\ell} \varphi_i \, \alpha_{in_1-1} \\
\Phi_{n_1} &= \varphi_1 \\
\Phi_j &= \sum_{i=2}^{\ell} \varphi_i \, \alpha_{ij} \text{ for } n_1 + 1 \leq j < n_2 \qquad (5.103) \\
\Phi_{n_2} &= \varphi_2 \text{ and so on.}
\end{aligned}$$

Using the property that all the coefficients α_{ij} which enter in the above formula are zero at λ_0, we see that restricting W_0 enough, and choosing $0 < r' \leq 1$ small enough, we have

$$W_{n_i}^{r'} \supset V_i^r \ , \text{ for } i = 1, \ldots, s,$$

for $W_i^{r'}$ defined as above. This is the property (ii), writing r' in place of r.

The property (i) for the factors H_i follows from the construction: we can take $H_1 = f_1, \ldots, H_{n_1-1} = f_{n_1-1}, H_{n_1} = \sum_{j \geq n_1} \alpha_{1j}(\lambda)f_j, H_{n_1+1} = f_{n_1+1}$, and so on.

The rest of the proof mimics the proof of part (i) of Theorem 4.9 for regular periodic orbits: we will prove that there are less than n_i roots of $\{\delta_\lambda(x) = 0\}$, for $\lambda \in W_{n_i}^r$, $i \leq s$ (restricting λ, x).

In order to prove it, we construct, by an algorithm of *"differentiation-division"*, a sequence of functions $\delta_0 = \delta, \delta_1, \ldots, \delta_{n_i}$ such that the last one is locally non-zero, and the bound for the number of roots will follow from a recurrent application of Rolle's theorem. For our present case, differentiation will produce non-bounded functions like x^{-k}. This is the reason that we have to arrange the expansion of δ_λ as in (5.102) with factor $H_i(x, \lambda)$ equivalent to $f_i(x, \lambda)$ for all λ (and not only for $\lambda = \lambda_0$ as in the smooth case), as we do not want to "leave behind us" some non-bounded function (with a small coefficient). A proof similar to the one we need was obtained by M. El Morsalani for general expansions: $\Sigma \alpha_{ij} (\lambda) x^i \omega^j + \psi_k$ of finite codimension [E2].

Here, we want to give a more direct proof which treats all terms in the sum (5.102) in the same way. For this, we need a more general algebra of admissible functions than the one introduced in Section 2.2. It will be the algebra of finite combinations of monomials:

$$x^{\ell + n\alpha_1} \omega^m \text{ with } \ell, n, m \in \mathbb{Z} \ (m \in \mathbb{Z} \text{ and not in } \mathbb{N}) \ \alpha_1 = \alpha_1(\lambda).$$

We order these monomials in a partial order as in Section 2.2. We introduce rational admissible functions $f = \dfrac{* \ x^{\ell + n\alpha_1} \ \omega^m + \cdots}{* \ 1 + \cdots\cdots}$ as in Section 2.2, with the same convention for $*$ (a continuous function of λ, which is non-zero at λ_0).

In order to make computations easier we extend the convention a little: by $+ \cdots$ we now denote a finite combination of monomials of order strictly greater than the *principal monomial* $x^{\ell + n\alpha_1} \omega^m$, plus some C^k function ψ which is k-flat at $x = 0$ (for the k chosen above).

For these new functions, the properties $a) - f)$ written in Section 2.2. remain valid. Moreover, we notice that $f)$ remains valid for ω^m, $m \neq 0$ as dominant term.

We now explain the algorithm. At each step we will produce a sequence of rational admissible functions. We just write their principal terms.

We begin with the sequence of N principal terms for the H_i:

$$1 \prec x\omega \prec x \prec x^2\omega^2 \prec x^2\omega \prec x^2 \prec x^3\omega^3 \prec x^3\omega^2 \prec x^3\omega \cdots$$

Differentiating once, we have the sequence of new principal terms:

$$0 \ , \ \omega \prec 1 \prec x\omega^2 \prec \ x\omega \prec x \prec x^2\omega^3 \prec x^2\omega^2 \prec x^2\omega \cdots$$

Next, dividing by the lowest term, we obtain

$$1 \prec \omega^{-1} \prec x\omega \prec x \prec x\omega^{-1} \prec x^2\omega^2 \prec x^2\omega \prec x^2 \cdots$$

which contains one term less than the initial one.

We now explain the induction step:

– assume we are given a sequence of rational admissible functions with a strictly increasing sequence of principal terms,

$$x^{j + n\alpha_1} \ \omega^m \ , \ \text{with } j \geq 1, \ n, \ m \in \mathbb{Z} \text{ or } j = 0 \text{ and } n = 0$$

(for instance ω^m for $m \in \mathbb{Z}$). We have to consider two different possibilities for the sequence of principal terms,

(i) First possibility:

$$1 \prec \omega^{-\ell_1} \prec \cdots \prec \omega^{-\ell_v} \prec \cdots \prec x^{j+n\alpha_1} \omega^m, \qquad (5.104)$$

with $1 \le \ell_1 < \ell_v$ $(v \ge 1)$ and $j \ge 1$, the same number n for all terms; the first function is 1.

Then a differentiation followed by the division by the first term of the sequence gives

$$1 \prec \cdots \prec \omega^{-(\ell_v - \ell_1)} \prec \cdots \prec x^{j+(n+1)\alpha_1} \omega^m. \qquad (5.105)$$

(ii) Second possibility:

$$1 \prec x^{j_0+n\alpha_1} \omega^{m_0} \prec \cdots \prec x^{j+n\alpha_1} \omega^m. \qquad (5.106)$$

Now we have no "pure" term in ω, $1 \le j_0 < j$, $n, m \in \mathbb{Z}$, the same n for each term and the first function is 1.

The operation of differentiation-division gives

$$1 \prec \cdots \prec x^{j-j_0+n\alpha_1} \omega^{m-m_0}. \qquad (5.107)$$

In both cases, we obtain a similar sequence to the initial one, but with one term less.

After j steps, the first j functions become zero and the $(j+1)^{th}$ one is transformed into 1.

We apply the operation of *differentiation-division* to the sequence $H_1, \ldots, H_N$ and by linear combination to $\delta_\lambda(x) = \sum\limits_{i=1}^{N} \Phi_i \, H_i$. After n_p steps, for $1 \le p \le s$, we obtain a final function,

$$\Delta_\lambda^p(x) = \Phi_{n_p}.H_{n_p}^p + \Phi_{n_p+1} \, H_{n_p+1}^p + \cdots + \Phi_N \, H_N^p, \qquad (5.108)$$

where $H_{n_p}^p \equiv 1$ and the principal term of each H_j^p for $j > n_p$ is of order strictly greater than 0 (order of 1). This means that $H_j^p(x, \lambda) \to 0$, for $(x, \lambda) \to (0, \lambda_0)$.

Taking the size of W_0 and x_0 small enough, we may suppose that $|\, H_j^p(x, \lambda) \,| < \dfrac{1}{Nr}$. Now take $\lambda \in W_{n_p}^r$,

$$|\, \Delta_\lambda^p(x) \,| \ge |\, \Phi_{n_p}(\lambda) \,| - \sum_{j \ge n_p+1} |\, \Phi_j(\lambda) \,| \,|\, H_j^p(x, \lambda) \,| \ge |\, \Phi_{n_p}(\lambda) \,| \; (1-r) \quad (5.109)$$

for $(x, \lambda) \in [0, x_0] \times W_0$.

Let $\Sigma = \{\Phi_1 = \cdots = \Phi_N = 0\}$, be the set of zeroes of the Bautin Ideal in W_0. We have

$$\Phi_{n_p}(\lambda) = 0, \; \lambda \in W_{n_p}^r \Rightarrow \lambda \in \Sigma \cap W_{n_p}^r.$$

This means that the function $\Delta_\lambda^p(x)$ has no zero on $]0, x_0[$, for $\lambda \in W_{n_p}^r - \Sigma$. But this function differs from δ_λ by n_p derivations and so, $\delta_\lambda(x)$ has no more than n_p isolated zeros on $]0, x_0[$ for $\lambda \in W_{n_p}^r$.

Applying this argument to each set $W_{n_p}^r$, $p = 1, \ldots, s$, we obtain the first part of Theorem 25.

Proof of part (ii) of Theorem 25.

The proof is very similar to the one given for the smooth case, in Theorem 4.2: if $\varphi_1, \ldots, \varphi_\ell$ are independent functions of λ, we can choose local coordinates $\lambda_1, \ldots, \lambda_\ell, \ldots, \lambda_\Lambda$ in the parameter space, with $\varphi_i = \lambda_i$, $i = 1, \ldots, \ell$ and $\lambda_0 = (0, \ldots, 0)$. Supposing that the system $\varphi_1, \ldots, \varphi_\ell$ is adapted at 0, we have

$$\delta_\lambda(x) = \sum_{i=1}^{\ell} \lambda_i \, h_i(x, \lambda),$$

with order $h_i(0) = n_i$, $n_1 < n_2 < \cdots < n_s < \infty$, $n_j = \infty$, if $j \geq s + 1$.

Restricting ourselves to the subfamily $\lambda_{s+1} = \cdots = \lambda_\ell = 0$, we write λ for $(\lambda_1, \ldots, \lambda_s)$: $\delta_\lambda(x) = \sum_{i=1}^{s} \lambda_i \, h_i(x, \lambda)$, with the orders as above. As in the regular case, the key point to observe is that the sequence $h_1(x, 0), \ldots, h_s(x, 0)$ is a regular Tchebychef family of germs at 0. This point was proved for these functions with principal term $x^i (Lnx)^j$, $i \geq j \geq 0$ by P. Joyal [J2]. The end of the proof is exactly the same as in the smooth case, and we conclude that in this case, $Cycl(x_\lambda, \Gamma) \geq s - 1$.

5.3.2   An example in quadratic vector fields

We finish this section with an application of Theorem 25 to quadratic vector fields. We return to the example introduced in 4.3.5.3. There we studied the 6-parameter family of Kaypten-Dulac (4.39) for the value $\lambda_0(\lambda_6 = -1, \lambda_1 = \cdots = \lambda_5 = 0)$.

For this parameter value, the vector field is Hamiltonian with Hamilton function $H(x, y) = \dfrac{1}{2} y^2 + \dfrac{1}{2} x^2 + \dfrac{1}{3} x^3$. We have computed the cyclicity of the smooth cycles of X_{λ_0}. Now, we want to study the saddle connection for $X_{\lambda_0} : \Gamma \subset \{H = \dfrac{1}{6}\}$ (see Figure 5.11).

We have seen that the Bautin Ideal at λ_0 is generated by $\lambda_1, \lambda_5, \lambda_2 \, \lambda_4$, so that

$$\delta_\lambda(x) = \varphi_1(\lambda) h_1(h, \lambda) + \varphi_2(\lambda) h_2(h, \lambda) + \varphi_3(\lambda) h_3(h, \lambda),$$

with $\varphi_1 = -2\lambda_1$, $\varphi_2 = -\lambda_5$, $\varphi_3 = -\dfrac{2}{33} \lambda_2 \lambda_4$, with

$$h_1(h, \lambda_0) = I_0(h), \quad h_2(h, \lambda_0) = I_1(h) \quad \text{and} \quad h_3(h, \lambda_0) = hI_0 - 3I_1.$$

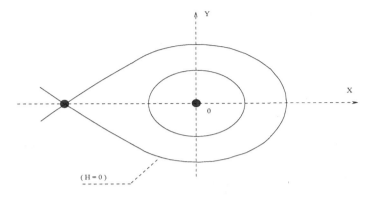

Figure 5.11

Let $u = \dfrac{1}{6} - h$ be a local parametrization of a transversal σ at Γ ($\Gamma \cap \sigma = \{u = 0\}$). It is easy to prove directly that any abelian integral $I(u) = \displaystyle\int_{\gamma_u} \omega$ for any algebraic 1-form ω has a Dulac series at $u = 0$:

$$I(u) = \gamma_0 + \gamma_1 u L n u + \gamma_2 u + o(u),$$

so that

$$
\begin{aligned}
h_1 &= I_0(u) = \alpha_0 + a_1 u L n u + a_2 u + 0(u) \\
h_2 &= I_1(u) = b_0 + b_1 u L n u + b_2 u + 0(u) \\
h_3 &= (\tfrac{1}{6}a_0 - 3b_0) + (\tfrac{1}{6}a_1 - 3b_1)u L n u + (\tfrac{1}{6}a_2 - 3b_2 - a_0)u + o(u).
\end{aligned}
$$

It follows from these expansions that functions h_1, h_2, h_3 are independent at order 2 (up to the "$uLnu$"-order) if and only if the 3×3 determinant

$$
\begin{vmatrix}
a_0 & a_1 & a_2 \\
b_0 & b_1 & b_2 \\
\tfrac{1}{6}a_0 - 3b_0 & \tfrac{1}{6}a_1 - 3b_1 & \tfrac{1}{6}a_2 - 3b_2\ a_0
\end{vmatrix}
\neq 0.
$$

This is equivalent to $a_0 \neq 0$ and

$$
\begin{vmatrix}
a_0 & a_1 \\
b_0 & b_1
\end{vmatrix}
\neq 0.
$$

The first condition is trivially fulfilled: a_0 is equal to the area of the disk bounded by Γ. To verify the second condition, we recall that the ratio $B_2(h) =$

$\dfrac{I_1}{I_0}(h)$ is equal to

$$B_2(h) = \frac{3}{4}\,\sqrt{2}B(\frac{\sqrt{3}}{8}\,(h - \frac{1}{12})) - \frac{1}{2},$$

where $B(h)$ is a solution of the Ricatti equation,

$$9(\frac{4}{27} - h^2)\,\frac{dB}{dh} = -7B^2 - 3hB + \frac{5}{3}. \qquad (5.110)$$

For $h = \frac{1}{6}$, $\frac{\sqrt{3}}{8}(h - \frac{1}{12}) = \frac{2}{3\sqrt{3}}$ and $B_2(u)$ has a Dulac series equivalent to the one of $B(\frac{2}{3\sqrt{3}} - u) = \overline{B}(u)$.

We have $B_2(h) = \dfrac{b_1}{b_0} + \dfrac{a_1 b_0 - a_1 b_1}{b_0^2}\,uLnu + o(uLnu)$, so that in order to prove that $a_1 b_0 - a_0 b_1 \neq 0$ it suffices to verify that

$$\overline{B}(u) = \gamma_0 + \gamma_1 uLnu + o(uLnu), \quad \text{with } \gamma_1 \neq 0. \qquad (5.111)$$

To compute γ_1, we write the Ricatti equation in $\overline{B}$, deduced from (5.110),

$$9u(\frac{4}{3\sqrt{3}} - u)\overline{B}' = 7\overline{B}^2 + 3(\frac{2}{3\sqrt{3}} - u)\overline{B} - \frac{5}{3} \qquad (5.112)$$

and substituting expansion (5.111) into (5.112) allows us to compute the coefficients γ_i by induction on i and to verify that $\gamma_1 \neq 0$.

Now, the independence of h_1, h_2, h_3 at order 2 is equivalent to $s_\delta(0) = 2$. We deduce from this that $Cycl(X_\lambda, \Gamma) \leq 2$ in the Kaypten-Dulac family.

As in Chapter 4, we can use Horozov's result for "generic hamiltonian": Horozov proved that the cyclicity of the saddle connection for such a generic hamiltonian is 2. The semi-continuity of cyclicity proved in Lemma 3 of Chapter 2, implies that $Cycl(X_\lambda, \Gamma) \geq 2$ and so, finally,

Theorem 26 *The cyclicity of the saddle connection for $\dot{x} = y$, $\dot{y} = x + x^2$ in the quadratic Kaypten-Dulac family is equal to 2.*

5.4   Unfoldings of elementary graphics

We now consider general elementary graphics with more than one vertex and some of the vertices may be semi-hyperbolic.

5.4.1   Hyperbolic graphic with 2 vertices

Suppose that X_λ is an unfolding in a neighborhood of a graphic Γ with two hyperbolic saddles p_1, p_2 as vertices at $\lambda_0 = 0$. Let $r_1(\lambda)$, $r_2(\lambda)$ be the hyperbolicity ratios of p_1, p_2, which are supposed to be hyperbolic saddles for each λ belonging to some neighborhood W_0 of $0 \in P$, where P is the parameter space. Let $r_1 = r_1(0)$, $r_2 = r_2(0)$. We have a first result, proved by Cherkas [C]. Here I follow a proof given in [DRR2].

Theorem 27 *Suppose the following generic conditions are fulfilled,*

$$r_1.r_2 \neq 1,$$

$r_1 \neq 1$, $r_2 \neq 1$. *Then* $Cycl\ (X_\lambda, \Gamma) \leq 2$.

Proof. We can suppose that $r_1.r_2 < 1$. (This means that the Poincaré map is expanding.) We can write the Poincaré map as a composition of four transitions: $P(x, \lambda_1) = G_2 \circ F_2 \circ G_1 \circ F_1(x_1, \lambda)$ (see Figure 5.12).

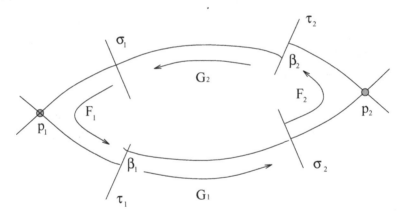

Figure 5.12

Here x_1 is a coordinate on the transversal segment σ_1. The transitions G_1, G_2 are differentiable and the transitions F_1, F_2 can be put into the Mourtada form:

$$F_1(x_1) = x_1^{r_1(\lambda)}(1 + \Phi_1(x_1, \lambda)) \qquad (5.113)$$
$$F_2(x_2) = x_2^{r_2(\lambda)}(1 + \Phi_2(x, \lambda)). \qquad (5.114)$$

We have to consider two different cases:

1) $r_1 < 1$, $r_2 < 1$. In this case $\dfrac{\partial F_1}{\partial x_1} \sim x_1^{r_1(\lambda)-1}$ and $\dfrac{\partial F_2}{\partial x_2} \sim x_2^{r_2(\lambda)-1}$. The derivative $\dfrac{\partial P}{\partial x_1}(x_1, \lambda) \to \infty$ when $(x, \lambda) \to (0,0)$, and we have at most one limit cycle near Γ, which is hyperbolic and expanding: $Cycl(X_\lambda, \Gamma) \leq 1$.

2) Up to the ordering of p_1, p_2 we can suppose that $r_1 > 1$ and $r_2 < 1$. In this case, instead of studying the solutions of $P(x_1, \lambda) = x_1$ we study the zeros of

$$H(x_1, \lambda) = G_1 \circ F_1(x_1, \lambda) - F_2^{-1} \circ G_2^{-1}(x, \lambda) = 0. \qquad (5.115)$$

We can write

$$G_1(y_1) = y_1 + \beta_1(\lambda), \ G_2^{-1}(x_1) = x_1 + \beta_2(\lambda),$$

$$(5.116)$$

$$F_1(x_1) = x_1^{r_1(\lambda)}(A_1(\lambda) + \Phi_1(x_1, \lambda)) \ and \ F_2^{-1}(x_2) = x_2^{s_2(\lambda)}(A_2(\lambda) + \Phi_2(x_1, \lambda)).$$

Here we put $s_2(\lambda) = \dfrac{1}{r_2(\lambda)}$. We have included the non-linear part of G_1, G_2^{-1} in F_1, F_2, using the fact that the smooth diffeomorphisms are in $\mathcal{D}$ (see Theorem 16); Φ_1, $\Phi_2 \in I$ and $A_1(0)$, $A_2(0) > 0$. We obtain

$$H(x, \lambda) = x^{r_1}(A_1 + \Phi_1(x, \lambda)) + \beta_1 - X^{s_2}(A_2 + \Phi_2(X, \lambda)). \qquad (5.117)$$

Here $x = x_1$, $X = x_1 + \beta_2$ and $\beta_1(\lambda)$, $\beta_2(\lambda)$ are the shift functions on the transversal segment τ_1 and τ_2 (see Figure 5.12).

A first differentiation of H gives

$$H'(x_1) = r_1 x^{r_1-1} (A + \Phi_1(x, \lambda)) - s_2 X^{s_2-1} (A_2 + \Phi_2(X, \lambda)), \qquad (5.118)$$

with ψ_1, $\psi_2 \in I$.

Hence, we can observe that $H' \neq 0$, when $X < x$ and x sufficiently small, i.e., there is at most one limit cycle for that position of the separatrices.

Zeros of $H'(x)$ are the same as zeros of

$$K(x) = x(B_1 + \xi_1(x, \lambda)) - \left(\frac{s_2}{r_1}\right)^{\frac{1}{r_1-1}} X^{\frac{r_2-1}{r_2-1}} (B_2 + \xi_2(X, \lambda)), \qquad (5.119)$$

with ξ_1, $\xi_2 \in I$ and $B_1(0)$, $B_2(0) > 0$.

We show that $K'(x)$ does not vanish in a neighborhood of $(x, y) = (0, 0)$ yielding a maximum of two zeros of $H(x)$. Indeed,

$$K'(x) = [B_1 + o(1)]-$$

$$\left[\left(\frac{s_2}{r_1}\right)^{\frac{1}{r_1-1}} \frac{s_2-1}{r_1-1} X^{\frac{r_2-r_1}{r_1-1}} (B_2 + \xi_2(X, \lambda))\right]. \qquad (5.120)$$

The first term stays bigger than a constant $M > 0$ while the second one goes to zero, as $(x, \lambda) \to (0, 0)$. $\qquad \square$

Generic two-parameter families are defined by the condition $\lambda \in \mathbb{R}^2 \to (\beta_1(\lambda), \beta_2(\lambda))$ has rank 2 at $0 \in \mathbb{R}^2$.

Such families are structurally stable and versal. Bifurcation diagrams can be found in [M3], [DRR2], for instance. The cyclicity of these versal families is 1 in the case $(r_1 - 1)(r_2 - 1) > 0$ and 2 in the case $(r_1 - 1)(r_2 - 1) < 0$.

A general study of the unfoldings for these 2-hyperbolic graphics was begun by Mourtada in [M3], [M4].

He classified the possibilities into the following cases:

$$
\begin{array}{lll}
(\mathcal{C}_1) & : & r_1 r_2 \neq 1 \\
(\mathcal{C}_2) & : & r_1 r_2 = 1 \text{ and } r_1 \notin \mathbb{Q} \\
(\mathcal{C}_3) & : & r_1 r_2 = 1 \text{ and } r_1 \in \mathbb{Q}
\end{array}
\tag{5.121}
$$

For the case $(\mathcal{C}_1)$ new problems arise when r_1 or $r_2 = 1$. We have to use the expansion in x, ω given in Proposition 12. The result, rather unexpected, is that the cyclicity is ≤ 2. For 3-parameter generic families (if $r_1(0) = 1$ for instance, we suppose $r_1'(0) \neq 0$), a bifurcation diagram is given in [M3] and [DRR2].

In [M3] Mourtada studied the case $(\mathcal{C}_2)$ under the hypothesis that the Poincaré map was not flat to identity. He proved that the absolute cyclicity of (X_0, Γ) is finite, i.e., there is a finite bound for the cyclicity depending only on (X_0, Γ) and not on the unfolding (X_λ, Γ). Moreover, he computed this cyclicity in terms of the resonant quantities at the saddle and of the Dulac series of the return map along Γ. The most interesting fact is that this absolute cyclicity does not depend just on the return map. In [M4] Mourtada considered the case $(\mathcal{C}_2)$ for an analytic unfolding (X_λ, Γ) such that X_0 has an identity return map. Using a division in the Bautin Ideal (like in Section 5.3 for the homoclinic connection), he proved that (X_λ, Γ) has finite cyclicity, of course not absolute but depending on the given unfolding (X_λ, Γ). As in Section 5.3 above he used an (x, ω) expansion of some shift function $\delta_\lambda(x)$, and divided it into the Bautin Ideal. Here the "compensator" $\omega(x, \lambda)$ is equal to $\omega(x, \lambda) = \dfrac{x^{-\alpha(\lambda)} - 1}{\alpha(\lambda)}$ where $\alpha(\lambda) = 1 - r_1(\lambda) r_2(\lambda)$.

The last case $(\mathcal{C}_3)$ was studied in [EM]. One of the difficulties in this case is that one has to work with two independent "compensators" $\omega_1(x, \lambda)$ and $\omega_2(x, \lambda)$ associated with each of the vertices p_1, p_2. In [EM], the authors used an idea already introduced in El Morsalani's thesis [E1], we eliminate one of the compensators by the "Khovanski-Moussu" procedure, explained below. They obtained partial results for this case. For instance, if the shift function $\delta(x, 0)$ is equivalent to $x^n \operatorname{Log}^m x$ for some $n, m \in \mathbb{N}$ and $m \neq 0$ (they say that $\delta(x, 0)$ has a logarithmic order), then the cyclicity of any unfolding (X_λ, Γ) is bounded by $\frac{1}{2} n(n + 5)$.

Hence, the proof of finite cyclicity of any analytic unfolding of a hyperbolic 2-graphic is almost complete. It remains to study the case $r_1(0) = r_2(\lambda)^{-1} \in \mathbb{Q}$ and $\delta(x, 0)$ is equivalent to ax^n for some non-zero constant a or is equal to identity.

Similar results were obtained by A. Jebrane and H. Zoladek [JZ], and A. Jebrane and A. Mourtada [JM] for the "figure height"-graphic.

5.4.2   Generic unfoldings of hyperbolic k-graphics

A hyperbolic k-graphic Γ is a hyperbolic graphic with k vertices: $p_1, \ldots, p_k$. Let X_λ be an unfolding of such a graphic defined for X_0 (see Figure 5.13). We label the vertices in circular order ($p_{k+1} = p_1$). Taking a transversal segment σ_i between the vertices p_i, p_{i+1}, we have a well-defined shift function $\beta_i(\lambda)$ on each σ_i, the difference between the first intersection of the unstable separatrix of p_i and the first intersection of the stable separatrix p_{i+1}.

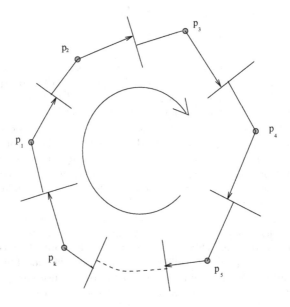

Figure 5.13

We consider generic k-parameter families. A first generic condition is that

(H1): $\lambda \to (\beta_1(\lambda), \ldots, \beta_k(\lambda))$ has rank k at $0 \in \mathbb{R}^k$. (Observe that $\beta_1(0) = \cdots = \beta_k(0) = 0$.)

Next, we suppose that the hyperbolic saddle points $p_1, \ldots, p_k$ remain fixed for $\forall \lambda \in W_0$ (some neighborhood of $0 \in \mathbb{R}^k$). Let $r_1(\lambda), \ldots, r_k(\lambda)$ be the hyperbolicity ratios.

Generic conditions of another type are put on (X_0, Γ) by imposing conditions on the hyperbolicity ratio s $r_i = r_i(0)$. These conditions are that Γ and any subgraphics which may be created by an unfolding have a "non-degenerate" return map:

(H2): For each subset $J \subset \{1, \ldots, k\} : \prod_{j \in J} r_j \neq 1$.

The conditions (H1), (H2) are the generic conditions introduced above for the case $k = 2$. Mourtada in [M1], [M2], proved a general result of finite cyclicity for generic unfoldings of hyperbolic k-graphics.

Theorem 28 *Let (X_0, Γ) be a generic smooth hyperbolic k-graphic (the generic conditions including the conditions (H2) among other explicit rational conditions on the ratios r_i). Let (X_λ) be a generic unfolding of (X_0, Γ), i.e., satisfying (H1). Then $\mathcal{C}ycl(X_\lambda, \Gamma)$ is finite.*

Moreover there exists a function $K(k) : \mathbb{N} \to \mathbb{N}$ such that

$$\mathcal{C}ycl(X_\lambda, \Gamma) < K(k)$$

(the bound is uniform, independent of the unfolding X_λ). $K(2) = 2$, $K(3) = 3$, $K(4) = 5$ and $K(k)$ is given inductively by a formula given in [M2].

Remark 30 *The cyclicity of (X_λ, Γ) is absolute in the sense that it depends only on (X_0, Γ). The genericity conditions in Theorem 28 are rational inequalities in $(r_1, \ldots, r_k)$. They define an open dense semi-algebraic subset U in $\mathbb{R}^k$ (the space for $(r_1, \ldots, r_k)$) and for each connected component of U there is a given cyclicity. This was done explicitly in Mourtada's thesis for $k = 3, 4$. In the case $k = 2$, we have two connected components, $U_1 = \{r_1 r_2 \neq 1, (r_1 - 1)(r_2 - 1) > 0\}$ with cyclicity 1 and $U_2 = \{r_1 r_2 \neq 1, (r_1 - 1)(r_2 - 1) < 0\}$ with cyclicity 2.*

A striking fact is that the cyclicity may be strictly bigger than the number of parameters, in contradiction to a previous conjecture of J. Sotomayor: for instance it may be equal to 5 for $k = 4$.

It is not possible to give here even a partial proof of this result (this proof in [M1], [M2] is more than 100 pages!). We just want to give a rough idea of some important steps.

We introduce $2k$ transversal segments σ_i, τ_i, $i = 1, \ldots, k$ (see Figure 5.14) and study the return map on σ_1.

To have a heuristic idea about what the return maps look like, suppose that each transition near a saddle is $x_i \to x_i^{r_i}$ (as if X_λ were linear near p_i) and that the regular map near τ_i and σ_{i+1} is a translation by β_i. (Of course we may suppose, under the assumed generic conditions, that $\lambda = (\beta_1, \ldots, \beta_k)$.)

Then, the return map P is equal to

$$P(x, \lambda) = \left(\cdots ((x_1^{r_1} + \beta_1)^{r_2} + \beta_2) \cdots \right)^{r_k} + \beta_k. \tag{5.122}$$

In a first part of his proof, in [M1], Mourtada proved that this normal form (5.122) is equivalent to P, up to some non-essential perturbation. To formulate this result more precisely, we have to generalize the property I^k introduced in

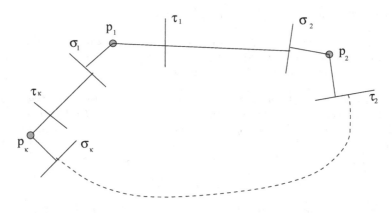

Figure 5.14

Section 5.1.3.3. Suppose that $\eta(\lambda)$ is some continuous function with $\eta(0) = 0$. We say that a function $f(x, \lambda)$ defined in a neighborhood of $(x, \lambda) = (0, 0)$ in the set $\{x \geq \eta(\lambda)\}$ is of "class I_λ^k for $(x - \eta(\lambda))$" if it is C^k, and verifies the properties i), ii) in the definition of the class I^k, with x replaced by $x - \eta(\lambda)$ and the limit $(x - \eta)^j \dfrac{j^j f}{\partial x^j} \to 0$ supposed for $(x, \lambda) \to (0, 0)$ (not uniformly in λ!) on its domain of definition.

We can define two natural functions $\rho(\lambda)$, $\eta(\lambda)$ in the following way:

1) $\rho(\lambda) = \text{Inf } \{x \in \sigma_1) \mid P(x, \lambda) : \sigma_1 \to \sigma \text{ is defined }\}$.

Of course, this means that the trajectory of $\rho(\lambda) \in \sigma_1$ tends towards some p_i and $\rho(\lambda)$ is the largest value of x with this property; $P(x, \lambda)$ is defined on $[\rho(\lambda), \varepsilon[$ for some fixed ε, and $\rho(\lambda)$ is a continuous function with $\rho(0) = 0$.

2) We can define a function $\widetilde{\eta}(\lambda)$ in a similar way as above for the trajectories of $-X_\lambda$, starting from τ_k. If $G_k(x, \lambda)$ is the transition map from τ_k to σ_1 we define $\eta(\lambda) = G_k(\widetilde{\eta}(\lambda), \lambda)$. It is a continuous function with $\eta(0) = 0$. Moreover, $\rho(\lambda) \geq 0$, $\rho(\lambda) \geq \eta(\lambda)$ and we may have $\eta(\lambda) < 0$.

Let $F_1, \ldots, F_k$ be the transition maps near the saddles, $F_i : \sigma_i \to \tau_i$ and $G_1, \ldots, G_k$ the regular transitions, $G_i : \tau_i \to \sigma_{i+1}$. For technical reasons, Mourtada preferred to replace the return $P(x, \lambda)$ by the difference

$$F_k \circ \cdots \circ G_2 \circ F_2 \circ G_1 \circ F_1 - G_k^{-1} = \Delta(x_1, \lambda). \qquad (5.123)$$

The roots of $\Delta(x_1, \lambda)$ on the domain $\mathcal{U} = \bigcup_{\lambda \in W_0}]\rho(\lambda), \varepsilon[\times \{\lambda\}$ correspond to limit cycles sufficiently near Γ.

It is now possible to give the first result of Mourtada.

Theorem 29 *(Reduction to the normal form [M1]). Let $\rho(\lambda)$, $\eta(\lambda)$ be the functions and $\mathcal{U}$ be the domain that we have introduced above.*

Then,

$$\Delta(x_1, \lambda) = \left(\cdots (x_1^{r_1} + b_1)^{r_2} + \cdots b_{k-1} \right)^{r_k} + b_k \varphi(x_1, \lambda). \qquad (5.124)$$

Here $b_1, \ldots, b_k$ are continuous functions of λ ($b_i(\lambda) = 0 \iff \beta_i(\lambda) = 0$); $\varphi(x_1, \lambda) = x_1(\alpha(\lambda) + f(x_1, \lambda))$, where $\alpha(\lambda)$ is positive and continuous on W_0 and $f(x, \lambda)$, defined on $V = \bigcup_{\lambda \in W_0}]\eta(\lambda), \varepsilon[\times \{\lambda\}$, is of class I_λ^k for $(x - \eta(\lambda))$ (see above).

Proof. The following gives a rough idea of the proof of Theorem 29: taking any coordinate y_1 on τ_1, Mourtada's form for F_1 is $F_1(x, \lambda) = x^{r_1} (A_1(\lambda) + \Phi_1(x, \lambda))$, with $A_1(0) > 0$. Now, it is possible to change the coordinate y_1 by a change $\bar{y}_1 = y_1(1 + \psi_1(x, \lambda))$, with ψ_1 of class I^k such that F_1 is given by $\bar{y}_1 = x^{r_1}$. And so on, we can change the coordinate on any transversal segment to reduce each regular map to a translation and each saddle transition to $x_i \to x_i^{r_i}$. Of course the change of coordinate may have some singularity at points other than 0: for this reason, we have to work with the larger class I_λ^k for some functions $x - u(\lambda)$. Also, the change of coordinates on transversal segments explains why the shift functions $\beta_i(\lambda)$ are replaced by the new functions $b_i(\lambda)$. At the end of the construction we have reduced the transition map $\sigma_1 \to \tau_k$ to its "normal form" and the transition G_k^{-1} to $\varphi - b_k$. $\qquad \square$

To obtain the computation of the number of roots of $\{\Delta(x_1, \lambda) = 0\}$ we have to show that the term φ plays no role in comparison with the principal normal part. Then the study would be reduced to the study of the roots of this principal part. This was done in [M2], which is devoted to the "finiteness algorithm" for the roots of $\{\Delta = 0\}$. I just indicate the first steps of this algorithm, in the case $k = 4$.

We define inductively,

$$h_j(x, \lambda) = [h_{j-1}(x, \lambda)]^{r_j} + b_j$$

$$\text{and } \Delta_\lambda(x) = \Delta(x, \lambda) = h_4 - x(1 + f). \qquad (5.125)$$

1) **First step:** (we eliminate b_4 by a differentiation in x)

$$\Delta_\lambda'(x) = *h_3^{r_4-1} h_2^{r_3-1} h_1^{r_2-1} h_0^{r_1-1} - (1 + f_{1,0}(x, \lambda)), \qquad (5.126)$$

$*$ is some constant and $f_{1,0}$ is a function in I_λ^k.

2) **Second step:** (we linearize in h_3 and eliminate it)

$$\Delta_\lambda'(x) = 0 \text{ is equivalent to}$$
$$\Delta_{1,\lambda}(x) = *(1 + f_{1,1})h_2^{z_3} h_1^{z_2} h_0^{z_1} - h_3 = 0, \qquad (5.127)$$

with $z_j = \dfrac{r_j - 1}{1 - r_4}$.

We can differentiate

$$\Delta'_{1,\lambda}(x) = *(1 + f_{2,0})h_2^{z_3-1} \ h_1^{z_2-1} \ h_0^{z_1-1}\Sigma \ - r_1 r_2 r_3 \ h_2^{r_3-1} \ h_1^{r_2-1} \ h_0^{r_1-1} \quad (5.128)$$

with

$$\Sigma = r_1 r_2 z_3 \ h_1^{r_2} \ h_0^{r_1} + r_1 z_2 h_2 h_0^{r_1} + z_1(1 + f_{2,1})h_2 h_1.$$

3) **Third step:** (we simplify $\Delta'_{1,\lambda}$ and derive)

Divide $\Delta'_{1,\lambda}$ by $h_2^{r_3-1} \ h_1^{r_2-1} \ h_0^{r_1-1}$. Then $\Delta'_{1,\lambda}$ has the same roots as

$$\Delta_{2,\lambda}(x) = *(1 + f_{2,0})h_2^{y_3} h_1^{y_2} h_0^{y_1} \Sigma - r_1 r_2 r_3 \quad \text{with} \quad y_j = z_j - r_j. \quad (5.129)$$

The derivative of $\Delta_{2,\lambda}$ has the same roots as

$$\Delta_{3,\lambda}(x) = Q^{2,2}(x,\lambda) = [h_2]^2 \ Q^{2,1} + h_2 \ S_2 Q^{1,1} + (S_1)^2 \ Q^{0,1}, \quad (5.130)$$

$$\text{with} \quad S_2 = h_1^{r_2} h_0^{r_1}, \quad S_1 = h_0^{r_1}. \quad (5.131)$$

$Q^{m,1}$ is a function which is homogeneous of degree m in (h_1, S_1), having as coefficient functions $Q^{m,0} = * + f_{i_m,j_m}$.

(Everywhere $*$ denotes a positive constant and $f_{i,j}$ a function of class I_λ^k.)

We stop this computation here. We have just presented the beginning of Mourtada's text to show that the first two steps are of similar nature: one linearizes the expression in the parameter b_4 and then b_3, and one gets rid off it by differentiation. Unfortunately this process stops at the third step, and the last expression is just quadratic in b_2. To go on, Mourtada uses a more sophisticated argument based on the fact that the expression is polynomial in the parameter with coefficients he can control inductively.

This change in the algorithm at the third step explains why the formula for $K(k)$ changes at $k = 4$:$K(2) = 2$, $K(3) = 3$, but $K(4) = 5$.

5.4.3   Generic elementary polycycles

In [IY3], Il'yashenko and Yakovenko obtained a general result of finite cyclicity for elementary graphics.

Theorem 30 *For any $n \in \mathbb{N}$, there exists a number $E(n) \in \mathbb{N}$ (called an "elementary bifurcation number") such that the cyclicity of any generic n-parameter unfolding of elementary graphic is bounded by $E(n)$.*

Remark 31 *This result seems to generalize the previous one of Mourtada who studied only generic n-parameter unfoldings of hyperbolic n-graphics (in the result of Il'Yashenko-Yakovenko the genericity assumption implies that the number k of vertices is less than the number n of parameters). However, Mourtada obtained an explicit bound in function of k. On the contrary, although the function $E(n)$*

is a "primitive recursive" function, i.e., computable by an algorithm, the practical computation of it remains an open question. Another interesting problem would be to have an explicit genericity condition in Theorem 30. It would be important to know these genericity conditions in order to apply the result in given bifurcation problems.

I recall that in Mourtada's Theorem 28, the genericity conditions are explicit rational inequalities in the hyperbolicity ratios r_i and also on the independence of the shift parameters β_i.

The greatest interest of Theorem 30 is the proof of the existence of $E(n)$. We can also introduce more particular bounds like $E(k,n)$ for the generic elementary graphics with k vertices and n parameters, $H(k,n)$, $H(n)$ for hyperbolic graphics. Of course $E(n) = Sup\ \{E(k,n) \mid k \leq n\}$, $H(n) = Sup\{H(k,n) \mid k \leq n\}$.

These numbers are known for small values of n (for instance $H(2) = 2$). A review of these results and also a description of all generic elementary unfoldings for $n \leq 3$ was made by Kotova and Stanzo [K-S].

Sketch of proof.

The complete proof given in [IY3], is rather long and profound. A good introduction to it and to related results was made in [IY2]. We just quote a sketch of the proof from this article (quoted in italics). Personal comments are added between quotation marks:

(The proof) consists of four principal steps:

1) $\mathcal{C}^k$-*smooth normalization of the family near each elementary singularity. The main tool here is provided by the classification theorems from [YI1]* ("see also Paragraph 1, [Bon]"). *The polynomial normal forms are integrable. We perform an explicit integration of normal forms in the class of Pfaffian functions introduced by Khovanskii [K] and show that the correspondence maps near each singular point in the normalization coordinates can be expressed through elementary transcendental functions which satisfy Pfaffian equations* ("see remark below"). *The degree and the total number of these equations can be estimated in terms of n.*

2) Algebraization *of the system of equations obtained at the previous step: the reduction procedure suggested in [K] allows to eliminate transcendental functions from the equations determining fixed points of the monodromy map. After this elimination, there appears a system of equations having the form of a chain map, a composition of a polynomial map and a jet of a generic smooth map.*

3) Gabrielov-type finiteness conditions *are established for a smooth map $F : \mathbb{R}^k \to \mathbb{R}^k$ to have a uniformly bounded number of regular preimages $\#F^{-1}(y)$ when the point y varies over a compact subset of $\mathbb{R}^k$. These conditions are automatically satisfied if a map F is real analytic. We introduce a topological complexity characteristic, the contiguity number, in terms of which an upper estimate for the number of preimage can be expressed.*

4) Thom-Boardman-type construction *allows us to prove that the above finiteness conditions can be expressed in terms of transversality of the jet extension of Γ to some semi-algebraic subsets of the jet space. Moreover, this construction can be generalized to cover chain maps of the form $P \circ (j^\ell F)$, where P in a polynomial, and $j^\ell F$ is the ℓ-jet extension of a generic smooth map. This is exactly the class of maps which appears after Khovanskii elimination procedure (step 2 above). The contiguity number of a chain map is expressed through the integer-valued data (degree of the polynomial P, order of the jet ℓ and dimension of the domain and target spaces).*

Remark 32 *The Khovanskii elimination procedure was first used for analytic vector fields by Moussu and Roche in [MoR] to prove the non-accumulation of limit cycles on elementary graphics (see Chapter 3) under the assumption that the normal form at each singular point in the graphic is convergent. For instance, we have seen above that the normal form at a singular hyperbolic saddle with resonance $p : q$ can be written*

$$\dot{x} = x \; , \; \dot{y} = \frac{1}{q}\left(-p + \sum_{i=0}^{\infty} \alpha_{i+1}(x^p y^q)\right)y.$$

Putting $u = x^p y^q$, this system is equivalent to

$$\dot{x} = x, \quad \dot{u} = P(u) = \sum_{i=1}^{\infty} \alpha_i u^i. \tag{5.132}$$

Under the assumption that P is convergent, we can obtain an analytic first integral for (5.132). If ω is the dual 1-form of (5.132), then $\omega = x\,du - P(u)dx = xP(u)dF$, with

$$F(x, y) = Q(u) - \, Log \; x \; , \; \text{where} \; \; Q(u) = \int_a^u \frac{ds}{P(s)}. \tag{5.133}$$

Now let $y = D(x)$ be the transition map from $\{y = 1\}$ to $\{x = 1\}$. Using the first integral F, we have

$$F(x, 1) = F(1, D(x)). \tag{5.134}$$

Hence $\{y = D(x)\}$ is the graph of an integral curve of the following 1-form Ω:

$$\Omega = \frac{\partial F}{\partial x}\,(x, 1)dx - \frac{\partial F}{\partial y}(1, y)dy.$$

Using expression (5.132) for F, it is easy to compute that Ω is equal (up to some analytic factor), to

$$\Omega = (px^p - P(x^p))P(y^q)dx - qy^{q-1}xP(x^p)dy. \tag{5.135}$$

Thus, $\{y = D(x)\}$ is the graph of an integral curve of the analytic 1-form Ω. We say that the function $D(x)$ satisfies the Pfaffian equation $\Omega = 0$. The *computation above is an example of the reduction procedure recalled in point 2 of the sketch of the proof.*

Another example of the same procedure was used by El Morsalani in [E1] to eliminate one compensator $\omega = \dfrac{x^{-\alpha} - 1}{\alpha}$ *from the equations. In this case, note that* $\dfrac{\partial \omega}{\partial x} = -x^{-1-\alpha} = -x^{-1}(\alpha\omega + 1)$, *so that* $x \to \omega(x, \alpha)$ *satisfies the Pfaffian equation* $x d\omega + (\alpha\omega + 1)dx = 0$.

In order to conclude this section, I want to mention partial, but interesting results about bifurcation studies and computation of cyclicity for elementary graphics:

- as recalled above Jebrane and Zoladek in [JZ] and others studied bifurcations of the "figure eight" (at a saddle point) of finite codimension. The study of the symmetrical case was made by Rousseau and Zoladek [RZ].

- El Morsalani in [E3] applied the reduction methods to graphics with two semi-hyperbolic vertices. It was proved in [IY2] that a figure made by two "opposite" semi-hyperbolic points of codimension 1 (forming a "lip-figure") can have an arbitrary "global cyclicity" even under generic assumptions (the cyclicity of each graphic in this lip-figure is less than 2).

- In a recent preprint [DER], Dumortier, El Morsalani and Rousseau proved that almost any non-trivial elementary graphic in the family introduced in Chapter 2, equivalent to the quadratic family, has finite cyclicity (in general ≤ 2). Here, a trivial elementary graphic is a graphic with identical return map. In [Z], Zoladek announced that the cyclicity of the trivial hyperbolic triangles in the quadratic vector family is equal to 3.

Chapter 6

Desingularization Theory and Bifurcation of Non-elementary Limit Periodic Sets

6.1 The use of rescaling formulas

In the study of the Bogdanov-Takens unfolding, we introduced in 4.3.5.2 the following formulas of rescaling in the phase-space and in the parameter space:

$$x = r^2 \, \bar{x}, \ y = r^3 \, \bar{y}, \ \mu = -r^4, \ \nu = r^2 \, \bar{\nu}.$$

Taking $(\bar{x}, \bar{y})$ in some compact disk $\overline{D}$ in $\mathbb{R}^2$, and $\bar{\nu}$ belonging to some closed interval K, the Bogdanov-Takens family is transformed into an r-perturbation of a generic Hamiltonian vector field X_0. Generic means here that the two singular points of X_0 are non-degenerate.

Of course, the whole operation reduces to studying the counter-image of the family $X_{\mu,\nu}$ (ignoring the extra-parameter λ) by the map $\Phi_r^S(\bar{x}, \bar{y}) = (r^2 \, \bar{x}, \ r^3 \, \bar{y})$, by making the change $Q^P(r, \bar{\nu}) = (\mu = r^4, \ \nu = r^2 \, \bar{\nu})$ in the parameter. The perturbation theories presented in Chapter 4 and Chapter 5 allow a complete study of the new family $\overline{X}_{r,\bar{\nu}} \, (\bar{x}, \bar{y})$ on the disk $\overline{D}$ and for $(r, \bar{\nu}) \in [0, U] \times K$, for some small $U > 0$. The study was made near regular cycles of the hamiltonian vector field X_0 in Chapter 4 and along the homoclinic loop in Chapter 5.

The maps Φ_r^s and the parameter change Φ^P are *singular* at $\{r = 0\}$. This point has the following two consequences:

1) The study of $\overline{X}$ covers only a conic sector $B = \Phi^P([0, U] \times K)$ in the parameter space, bounded by a curve having a quadratic contact with the 0ν-axis (see Figure 6.1). Therefore, in order to complete the study, we have to investigate the family $X_{\mu,\nu}$ in a complement of B in some neighborhood of the origin in the parameter space (see [B], [RW] for details).

2) For parameter $(\mu, \nu) \in B$, the study of $\overline{X}_{r,\bar{\nu}}$, on $\overline{D}$ gives the phase portrait of $X_{\mu,\nu}$ on the disk $D_r = \Phi_r^P(\overline{D})$ and the diameter of D_r tends to zero when

151

$r \to 0$. But, we need to obtain the phase portrait on a fixed disk D, surrounding the origin in the phase space.

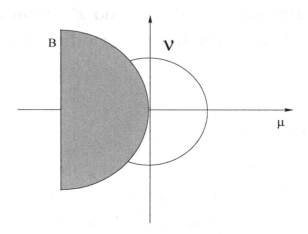

Figure 6.1

In the case of the Bogdanov-Takens unfolding, it is easy to choose a disk D such that no singularities exist in $D \backslash D_r$. Then, as a consequence of the Poincaré-Bendixson theory, we can prove that the vector field $X_{\mu,\nu} \mid D_r$ is topologically equivalent to $X_{\mu,\nu} \mid D$.

This difficulty is always present when one uses rescaling formulas in the study of unfoldings.

In general, rescaling a formula for an unfolding (X_λ) at $0 \in \mathbb{R}^p$ and at the origin in the parameter space $\mathbb{R}^k$ are the formulas:

$$\begin{cases} (x,y) &= \Phi_r^S(\bar{x}) = (r^{\alpha_1}\,\bar{x}, r^{\alpha_2}\,\bar{y}) \\ \mu &= \Phi^P\,(\bar{\mu}, r) = (r^{\beta_i}\,\bar{\mu})_{i=1,\ldots,k}, \end{cases} \tag{6.1}$$

where $\bar{\mu} = (\bar{\mu}_1, \ldots, \bar{\mu}_k)$.

Here we have $\lambda = (\mu, \Lambda)$. Notice that in general only some parameters μ_i are rescaled (for instance, in the Bogdanov-Takens family we have rescaled the parameters μ, ν and not the components of λ).

The coefficients α_i, β_j are chosen by considerations of quasi-homogeneity. For instance, we can choose the α_i, looking at the Newton diagram for the vector field X_0 ($\lambda = 0$). Next, we can choose the coefficients β_j in such a way that we have as many monomials as possible of the lowest degree of quasi-homogeneity in the family: this is equivalent to taking a face of the Newton diagram for the Taylor series of the family expanded in all variables x and μ. Let us consider

$$\left(\Phi_r^S\right)_*^{-1}(X_{\Phi^P}) = r^\delta\,\overline{X}_{r,\bar{\mu},\lambda}\,(\bar{x}) \tag{6.2}$$

where δ is as big as possible. Now, when *rescaling the family* we take $\bar{x}$ belonging to some compact domain $\overline{D}$, $\bar{\mu}$ belonging to the unit sphere S^{k-1} and $r \in \mathbb{R}^+$.

This choice covers a neighborhood in the parameter space (contrary to the restricted choice recalled above for the Bogdanov-Takens family), but the problem remains that the disk $D_r = \Phi_r^S(\overline{D})$ shrinks when $r \to 0$. This fact has several drawbacks. Some non-trivial bifurcation phenomena could happen in the region $D \backslash D_r$ and we have to study them or to justify that $\overline{X}_{|D_r}$ is equivalent to $\overline{X}_{|D}$. But the problem is even more serious when we use the rescaling formulas at a vertex of a limit periodic set that we want to study.

Consider for instance a graphic Γ of X_0 with the point p as a vertex. Then, the trajectories γ_1, with $\alpha(\gamma_1) = p$ and γ_2, with $\omega(\gamma_2) = p$ belong to Γ, and to study the unfolding (X_λ, Γ) we have to look at the transition map $T_\lambda : \sigma_1 \to \sigma_2$ near p, where σ_1, σ_2 are transversal segments to γ_1, γ_2, taken in ∂D. Here, D is a fixed neighborhood of $0 \in \mathbb{R}^2$ (see Figure 6.2). But a study by rescaling just allows us to study the transition maps between transversal segments taken in the boundary of D_r, whose distance to the origin tends to zero when $r \to 0$. We are left with an unsolved singular limit problem.

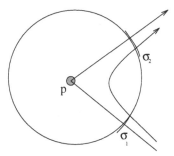

Figure 6.2

To overcome this difficulty, the idea is to consider the rescaling formulas as a chart in a *global generalized blowing-up*. The global blowing-up is defined by the map $\Phi(\bar{x}, \bar{\mu}, r, \Lambda) = (x, \mu, \Lambda)$ given by the rescaling formulas (6.1) (i.e., $\Phi = (\Phi_r^S, \Phi^P, \Lambda)$), when one takes $(\bar{x}, \bar{\mu}) \in S^{k+1}$, $r \in \mathbb{R}^*$, $\Lambda \in \mathbb{R}^{p-k}$. We see that the domain of the rescaling $\overline{D} \times S^{k-1}$ is homeomorphic to a part of S^{k+1}. In fact, corresponding to the decomposition $\mathbb{R}^{k+2} = \mathbb{R}^2 \times \mathbb{R}^k$, we have a related topological decomposition of S^{k+1},

$$S^{k+1} = \overline{D} \times S^{k-1} \ \cup \ S^1 \times D^k. \tag{6.3}$$

In this topological decomposition, we replace the "round sphere" by $\partial(\overline{D} \times D^k)$, where $D^k \subset \mathbb{R}^k$ is the disk centered at $0 \in \mathbb{R}^k$.

We consider the family X_λ as a vector field X, defined in $\mathbb{R}^{2+p}$.

We suppose that $X_\lambda(0) = 0$, for $\lambda = (0, \Lambda)$; so that there exists, in general, a smooth vector field $\widehat{X}$ such that $\Phi_*(\widehat{X}) = X$. This is the case for homogeneous blow-up, $\alpha_i = \beta_i = 1$ (if not, $\widehat{X}$ is smooth after multiplication by r^{δ_0}, for some $\delta_0 \in \mathbb{Z}$). In any case, there exists a bigger δ such that $\dfrac{1}{r^\delta} \widehat{X} = \overline{X}$ is a smooth vector field. This vector field $\overline{X}$ will be called the *desingularized vector field* (by the blow-up Φ). Of course $\overline{X}$ is not identically zero along the critical locus $S^{k+1} \times \{0\} \times \mathbb{R}^{p-k}$ and we can expect that $\overline{X}$ has simpler singularities than X.

Taking $r \in [0, U]$ for some $U > 0$, Λ belonging to some neighborhood W of $0 \in \mathbb{R}^{p-k}$, $(\bar{x}, \bar{\mu}) \in S^{k+1}$, we cover a whole neighborhood of $0 \in \mathbb{R}^{p+k}$ in the (x, y, λ)-space. A new difficulty is that $\overline{X}$ is *no longer a family*: it remains a family in the old *rescaling domain* (on $\overline{D}$, with parameters $(\bar{\mu}, r, \Lambda) \in S^{k-1} \times [0, U] \times W$), but not in the other part $S^1 \times D^k \times [0, U] \times W$, that we will call the *phase space domain* (see Figure 6.3).

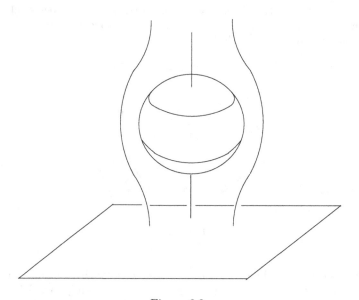

Figure 6.3

For practical computation, it is preferable in general to cover the two domains by an atlas of charts (we call them rescaling directional and phase-space directional charts). For instance, we can replace $(\bar{x}, \bar{y}) \in S^1$ by an atlas of directional charts: $\bar{x} \in K$ (some interval) $\bar{y} = \pm 1$, $\bar{x} = \pm 1$, $\bar{y} \in K$.

For such choices of $\bar{x}, \bar{y}$, the map Φ gives a vector field X_C on each chart C which differs from the global one $\overline{X}$ (obtained by taking $(\bar{x}, \bar{\mu}) \in S^{k+1}$) by a positive smooth multiplicative function.

Hence we replace $\overline{X}$ by the singular foliation it defines. We will call it below a *local vector field* to distinguish it from the two-dimensional foliation produced by the blowing-up of the fibration on the parameter space.

This global blowing-up was introduced in [R6] and the method was systematized in a work in collaboration with Z. Denkowska [DeR]. Next, some applications have been developed concerning particular unfolding questions: nilpotent focus point of codimension 3 [DR2], and Van der Pol equation [DR3]. In these notes we want to return to the first example treated in [R6], the cuspidal loop, because it is the simplest example of a non-elementary graphic and also because some progress was made recently. These new results will appear in the paper [DRS3] and will be presented here in the next section. In the last section, we want to explain how the global blowing-up could enter into the theory of desingularization of families, along the lines presented in [DeR], then compare it with the theory of Trifonov [Tr]. Finally, we point out that this subject has not yet reached its final form: the desingularization theory remains to be developed and we want to present some conjectures and some ideas on how to attack them.

6.2 Desingularization of unfoldings of cuspidal loops

A cuspidal loop is a singular cycle consisting of a cusp point p and a connection Γ between the two branches of the cusp. We want to study generic unfoldings of such a cuspidal loop. A first generic condition is that the cusp point is a codimension 2 singularity, i.e.,

$$j^2 \, X_0(p) \sim y \, \frac{\partial}{\partial x} + \left(x^2 + \varepsilon xy\right) \frac{\partial}{\partial y} \quad \text{with } \varepsilon = \pm 1. \tag{6.4}$$

The connection adds an extra condition so that it is natural to study generic 3-parameter unfoldings of (X_0, Γ).

Consider a segment Σ' transverse to the connection (see Figure 6.4). Let $P : \Sigma \to \Sigma'$ ($\Sigma \subset \Sigma'$, a neighborhood of $q = \Sigma' \cap \Gamma$), be the Poincaré map. *It is a* C^1-*map* (see below) and we require that $\gamma = P'(q) \neq 1$. Replacing X_0 by $-X_0$ we can suppose that $\gamma < 1$ (Γ is an attracting cycle). This is the generic condition (H1) for X_0.

Let us now consider a 3-parameter unfolding X_λ of X_0 near Γ for λ close to $0 \in \mathbb{R}^3$. As we have seen in Chapter 1, we can choose coordinates (x, y) near p, with $p = (0,0)$ such that in a neighborhood of p, X_λ is C^∞-equivalent to

$$\begin{cases} \dot{x} &= y \\ \dot{y} &= x^2 + \mu(\lambda) + y(\nu(\lambda) + \varepsilon x + x^2 h(x, \lambda)) + y^2 Q(x, y, \lambda), \end{cases} \tag{6.5}$$

where h and Q are C^∞ functions.

The second generic condition (H2) for the family is that *the map:* $\lambda \to (\mu(\lambda), \nu(\lambda))$ *is of rank 2 at* $\lambda = 0$.

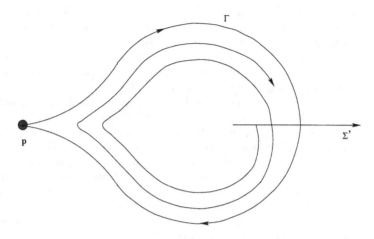

Figure 6.4

That is, we suppose that the system (X_λ) has a cusp at 0, for λ belonging to a regular line L passing through the origin in the parameter space $\mathbb{R}^3$.

For $\lambda \in L$, it makes sense to define a *shift map* between the two separatrices of the cusp. If Σ' is oriented as in Figure 6.4, and $u(\lambda)$, $s(\lambda)$ are the intersections of the unstable and the stable separatrix of p with Σ', respectively, we define $\sigma(\lambda) = u(\lambda) - s(\lambda)$.

Using the desingularization of p by the quasi-homogeneous blowing-up $x = r^2 \, \bar{x}$, $y = r^3 \, \bar{y}$, it is possible to show that σ is a C^∞ function of L.

The third generic condition (H3) for X_λ is that *the map $\lambda \in L \to \sigma(\lambda)$ has a non-zero derivative at $\lambda = 0$.* If we take *any* C^∞ extension of σ in a neighborhood of $0 \in \mathbb{R}^3$, conditions (H2) and (H3) imply that the map $\lambda \to (\mu(\lambda), \nu(\lambda), \sigma(\lambda))$ is of rank 3 at $0 \in \mathbb{R}^3$. Hence, up to a diffeomorphic change of parameter, we can suppose that $\lambda = (\mu, \nu, \sigma)$.

Definition 31 *In the following text, generic 3-parameter unfolding of a cuspidal loop will mean an unfolding which satisfies the three generic conditions (H1), (H2) and (H3).*

Knowing the codimension 2 phenomena and using some heuristic arguments (like the famous "simplicity principle") it is not too hard *to predict* the possible bifurcation diagrams.

Two different diagrams can be so produced, depending on the sign $\pm$ in the Bogdanov-Takens bifurcation. We have presented these diagrams in Figure 6.5 and Figure 6.6.

It is rather easy to prove the occurrence and genericity of the different saddle connections (lines L_r, L_ℓ, L_i, L_s). The difficult part of the proof deals with the limit cycles. The fact that the small limit cycle that appears in the Bogdanov-Takens bifurcation is expanding in the case $\varepsilon = 1$ induces a slightly more complicated bifurcation diagram, exhibiting four limit cycles.

Remark 33 *1) This number 4 is not without importance. In codimension 1 and 2 in the plane, any generic limit periodic set generates a certain number of limit cycles bounded by the codimension. Here, in generic 3-parameter unfoldings of the codimension 3 cuspidal loop, one may generate 4 limit cycles. It is a quite unexpected phenomenon, similar to the one observed by Mourtada for generic hyperbolic polycycles of codimension 4, which generate 5 limit cycles (see Chapter 5 above).*

2) Figures 6.5 and 6.6 represent intersections of the cone-like 2-dimension bifurcations set with a 2-sphere S^2 centered at $0 \in \mathbb{R}^3$. A point on S^2 was removed in order to present the bifurcation diagram in a plane.

In fact, the most interesting part of the bifurcation set is situated in a small cylinder $(\sigma \simeq 0)$. Indeed, for a fixed non-zero value $\sigma(\sigma > 0$ or $\sigma < 0)$, it is clear that we can only expect to find the bifurcation diagram of the Bogdanov-Takens bifurcation (for $(\mu, \nu) \sim (0,0)$). Therefore, for the study of the bifurcation set, it might be more natural to intersect it with a "cylinder box", $\{| \sigma | \leq S, x^2 + y^2 \leq r^2\}$.

The bifurcation diagrams in Figures 6.5, 6.6, which summarize the results about bifurcations of the generic unfolding remain conjectural. What precisely will be proved in [DRS3] is that the study can be reduced to the properties of some function:

Theorem 31 *The diagrams in Figures 6.5, 6.6, are implied by the property of monotonicity (M) of a "transition time" function $t_\varphi(x)$. (We define $t_\varphi(x)$ and the property (M) below in Subsection 6.2.3.)*

Remark 34 *1) All phenomena of bifurcations concerning saddle connections are easy to obtain as we said before. The reduction presented below permits us to study and check any codimension 2 phenomenon, expect the occurrence of the triple limit cycle (TC), without the use of property (M). In fact, property (M) is only needed to justify the results about limit cycles, number, cyclicity, and existence of bifurcations for them: TC, DC^{out} DC^{in} (see Figures 6.5, 6.6).*

2) C. Simó has obtained good "numerical evidence" for the property (M).

The next subsections are devoted to the presentation of the ingredients of the proof of Theorem 31. Details will appear in [DRS3]. These ingredients are needed to overcome the problem that separatrices of the cusp suddenly lose any geometrical meaning when $\mu > 0$, while on the region $\{\mu < 0\}$ they turn into separatrices of a saddle point, but in a non-differentiable way.

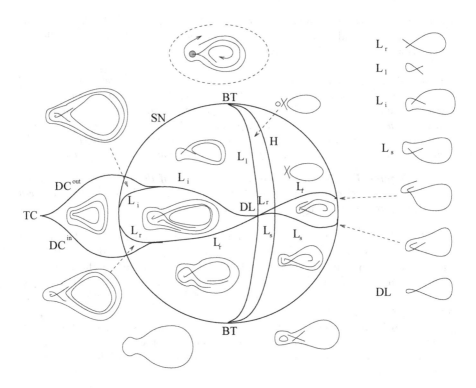

Figure 6.5 Case $\varepsilon = -1$ for $-X$

6.2.1 Global blowing-up of the cusp unfolding

At the point $(x, y, \mu, \nu) = (0, 0, 0, 0)$, we make the following global blowing-up of the family,

$$\begin{cases} \dot{x} & = & y \\ \dot{y} & = & x^2 + \mu + y(\nu + \varepsilon_1\, x + x^2\, h) + y^2\, Q \end{cases} \tag{6.6}$$

where $\varepsilon_1 = \pm 1$, $h(x, \lambda)$, $Q(x, y, \lambda)$ are $\mathcal{C}^\infty$-functions and $\lambda = (\mu, \nu, \sigma)$.

The blowing-up formulas are

$$\begin{cases} x & = & r^2\, \bar{x} \\ y & = & r^3\, \bar{y} \\ \mu & = & r^4\, v^4 \cos\, \varphi \\ \nu & = & r\, v \sin\, \varphi \\ \sigma & = & \sigma, \end{cases} \tag{6.7}$$

with $\bar{x}^2 + \bar{y}^2 + v^2 = 1$ (or $(\bar{x}, \bar{y}, v) \in \partial B$, where B is some "box" homeomorphic to D^3).

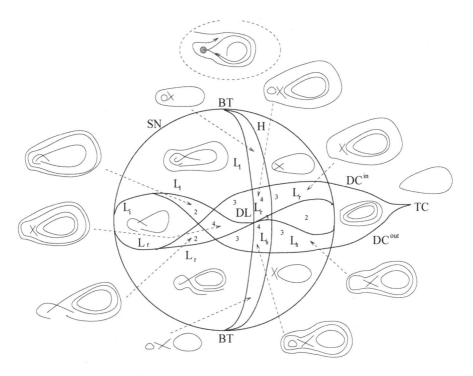

Figure 6.6 Case $\varepsilon = 1$

Thus, (6.7) defines a mapping $\Phi : S^2 \times \mathbb{R}^+ \times S^1 \times \mathbb{R} \to \mathbb{R}^5$, $\Phi : ((\bar{x}, \bar{y}, v), r, \varphi, \sigma) \to (x, y, \mu, \nu, \sigma)$.

Remark 35 *Here, we compose the global blowing-up with "small-parameter" r, as described in Paragraph 1 with a polar type blowing-up $\bar{\mu} = v^4 . \cos\varphi$, $\bar{\nu} = v \sin\varphi$. The critical locus is not S^3 (for any fixed σ) but $S^2 \times S^1$, and we pass from one to the other by the branched covering map $S^2 \times S^1 \to S^3$ which is the blowing-up map of S^3 along one circle.*

We now have a $\mathcal{C}^\infty$ vector field $\overline{X} = \frac{1}{r} \widehat{X}$ with $\Phi_\lambda(\widehat{X}) = X$ and X is the family (X_λ), which we consider as a vector field in $\mathbb{R}^5$.

For each constant value of (φ, σ), $\overline{X}$ induces a 3-dimensional vector field $\overline{X}_{(\varphi,\sigma)}$ defined on $S^2 \times \mathbb{R}^+$ near $S^2 \times \{0\}$. Hence, the global blowing-up map Φ transforms our 3-parameter family into a (φ, σ)-family $\overline{X}_{(\varphi,\sigma)}$ which we want to describe next. Several facts can be observed which makes this description easier:

1) The set $\{v = 0\}$ is invariant and the restriction of the map Φ to it is the usual quasi-homogeneous blowing-up of the cusp singularity ($\mu = \nu = 0$), studied in Chapter 3 (see Figure 6.7).

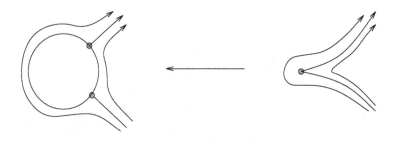

Figure 6.7

2) It suffices to study $\overline{X}_{(\varphi,\sigma)}$ on $\{v \geq 0\}$ in order to get complete information on the (x, y, λ)-space near $0 \in \mathbb{R}^5$.

3) The vector field $\overline{X}_{(\varphi,\sigma)}$ leaves invariant the foliation given by $\{rv\} = \mathcal{C}^{st}$, for each (φ, σ). This is the foliation obtained by the blow-up of the foliation of $\mathbb{R}^5$ in the parameter space (μ, ν, σ). The leaf for $\{rv = u\}$ with $u > 0$ is a regular 2-manifold, and for $\{rv = 0\}$ we get a stratified set as in figure 6.8, with 2 strata of dim 2: $\widehat{F} \simeq S^1 \times \mathbb{R}$ (the blowing-up of the fiber $\mu, \nu = 0$) and the hemisphere $D_{\varphi,\sigma} = \{\bar{x}^2 + \bar{y}^2 + \bar{v}^2 = 1, \bar{v} \geq 0\}$ contained in the critical locus.

We already know the behaviour of $\overline{X}_{(\varphi,\sigma)}$ on $\widehat{F} = \{v = 0\}$, and in order to obtain the behaviour of $\overline{X}_{(\varphi,\sigma)}$ on $D_{\varphi,\sigma} = \{r = 0\}$, we have to study $\overline{X}_{(\varphi,\sigma)}$ on the half-sphere $D_{\varphi,\sigma}$.

Near $\{v = 0\}$ we use *phase directional rescaling*. This means that in (6.8) we take $\bar{x}^2 + \bar{y}^2 = 1$ and v near 0. (It is better to use subcharts by taking $\bar{x} = \pm 1$, resp. $\bar{y} = \pm 1$.)

Taking $\bar{x} = 1$, we obtain the phase portrait of the vector field $\overline{X}^x$ represented in figure 6.9, for each value (φ, σ).

The singularities at p_1 and p_2 are hyperbolic saddles; the eigenvalues at p_2 are $-6, -1, 1$ (-6 along the $\bar{y}$-axis). At p_1 we have eigenvalues equal to $6, 1, -1$ (with 6 along the $\bar{y}$-axis).

The relation between the eigenvalues at p_1 and p_2 is not a coincidence since in fact the two points are "the same". To see this, we use $\bar{y} = 1$ (instead of $\bar{x} = 1$), giving a vector field $\overline{X}^y$. As in the global blowing-up Φ, $y = r^3\,\bar{y}$, the chart $\bar{y} = 1$ includes $\bar{y} = -1$ by changing $(\bar{x}, r, v, \lambda, t) \rightarrow (\bar{x}, -r, -v, \lambda, -t)$. We obtain a singular point p on the $0\bar{x}$ axis (see Figure 6.10). The study of $\overline{X}^y$ at p includes up to a $\mathcal{C}^\infty$ equivalence, the study of $\overline{X}^x$ near p_2 on $\{v \geq 0, r \geq 0\}$, and the situation near p_1, on $\{v \leq 0, r \leq 0\}$ (if we reverse time).

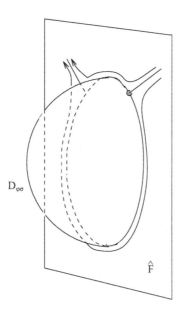

Figure 6.8

In order to complete the picture on the 2-sphere, we use the *"family rescaling"* (i.e., the usual rescaling). In (6.7), we take $v = 1$, leading to a (r, φ, σ) family of 2-dimensional vector fields. On $\{u = 0\}$ this gives the following family of vector fields:

$$X_\varphi \begin{cases} \dot{\bar{x}} &= \bar{y} \\ \dot{\bar{y}} &= \bar{x}^2 + \cos\varphi + \bar{y}\sin\varphi. \end{cases} \tag{6.8}$$

This is a very simple φ-parameter quadratic family of vector fields X_φ (it is independent of σ); we can draw the picture on the Poincaré disk. At ∞ (the boundary of the Poincaré disk) we add the knowledge that we obtained by phase directional rescaling. Notice that it corresponds to the use of the quasi-homogeneous compactification: $x = \dfrac{\cos\theta}{v^2}$, $y = \dfrac{\sin\theta}{v^3}$ ($\{v = 0\} = \infty$) (see Figure 6.11).

A set is a limit periodic set of $\overline{X}$ if it is the limit (in the sense of Hausdorff distance) of a sequence of limit cycles in the blowing-up space. All of them are in $\{\sigma = 0\}$ inside $\widehat{F} \cup D_{\varphi,0}$, for some φ. They are closed curves made by the connection $\widehat{\Gamma}$ from p_1 to p_2 in $\widehat{F}$ (coming from Γ) and a connection between p_2 and p_1 in $D_{\varphi,0}$. They are all "elementary" in the sense that they just include the two singular hyperbolic points p_1, p_2 and perhaps a saddle point, as singular point. See Figure 6.12 for the different possibilities of limit periodic sets of $\overline{X}$.

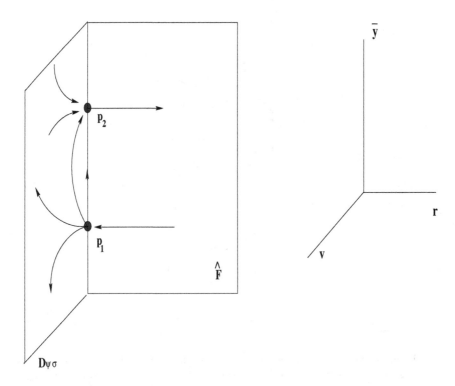

Figure 6.9

6.2.2   Asymptotic form for the shift map equation

Let Σ_1, Σ_2 be two transversal sections to orbits, contained in $\{v = v_0\}$, near the points p_1, p_2 respectively. These sections are rectangles parametrized by (θ, r). We consider the transition maps for the flow of $\pm \overline{X}_{\varphi,\sigma}$ from Σ_1 to Σ_2. First let $G_{\varphi,\sigma}$: $\Sigma_1 \to \Sigma_2$ be the map obtained by following the flow of $\overline{X}_{\varphi,\sigma}$ ($G_{\varphi,\sigma}$ is the transition map near the disk $D_{\varphi,\sigma}$), and next we let $R_{\varphi,\sigma}$ be the map obtained by following the flow of $-\overline{X}_{\varphi,\sigma}$ ($R_{\varphi,\sigma}$ is the transition map near $\widehat{F}$).

Limit cycles cutting Σ_1 correspond to the solution of the shift map equation,

$$\Delta_{\varphi,\sigma} = G_{\varphi,\sigma} - R_{\varphi,\sigma} = 0. \tag{6.9}$$

In order to obtain all the information about the bifurcation diagram it will be sufficient to study the equation $\{\Delta_{\varphi,\sigma} = 0\}$, when v_0 is chosen small enough. Of course, the rest of the study will be completed by considering a similar shift map for sections Σ_1', Σ_2' chosen transversally to $\widehat{F}$, but it is easy to verify that any limit cycle cutting Σ_1' (small enough) must be a hyperbolic attracting cycle: we can

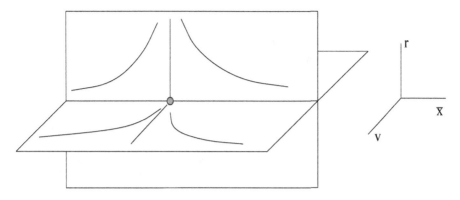

Figure 6.10

prove that the integral of the divergence is always negative. Hence, no bifurcation can happen at such a limit cycle.

In order to obtain an interesting asymptotic form for $\Delta_{\varphi,\sigma}$, when the critical parameter $u = rv$ tends to 0, we introduce a "good normal form" coordinate on Σ_1, Σ_2, coming from some normal form coordinates at the singular point p which gives birth simultaneously to p_1 and p_2.

Recall that $\overline{X}_{\varphi,\sigma}$ has resonant eigenvalues $-1, -6, 1$ at p and moreover that the function $u = rv$ is a first integral. It follows from this that $\overline{X}_{\varphi,\sigma}$ is *linear in r and v*, up a multiplicative function.

It is then possible to reduce $\overline{X}_{\varphi,\sigma}$ to a $\mathcal{C}^\infty$ normal formal by a $\mathcal{C}^\infty$ diffeomorphism $((\varphi,\sigma)$-dependant), in a neighborhood of p. In fact the two variables r, v are preserved by the diffeomorphism (see details in [DRS3]). Let us formulate this special normal form theorem:

Theorem 32 *Let X_λ be a parametrized $\mathcal{C}^\infty$ vector-field family in a neighborhood of $0 \in \mathbb{R}^3$ ($\lambda \in \mathbb{R}^p$, near $0 \in \mathbb{R}^p$).*

Let the differential equation for X_λ be given by

$$\dot{r} = -r, \ \dot{v} = v \ \text{ and } \ \dot{\Omega} = 6\Omega + \widetilde{G} \, (\Omega, \ r, \ v, \ \lambda).$$

Then, for any $k \in \mathbb{N}$, there exist a constant $N(k)$ and a $\mathcal{C}^k$-family of diffeomorphisms $H_\lambda(r, v, \Omega) = (r, v, h_\lambda(r, v, \Omega))$ which brings X_λ to an $N(k)$-normal form:

$$\dot{r} = -r, \ \dot{v} = v, \ \dot{\Omega} = 6\Omega + \widetilde{G}_\lambda^N(r, v, \Omega),$$

where $\widetilde{G}$ is a resonant polynomial of degree $N = N(k)$:

$$\widetilde{G}_\lambda^N(r, \ v, \ \Omega) = \sum_{\substack{i + j + \ell \leq N \\ 6i - j + \ell = 6}} \alpha_{ij\ell} \, (\lambda)\Omega^i \, r^j \, v^\ell.$$

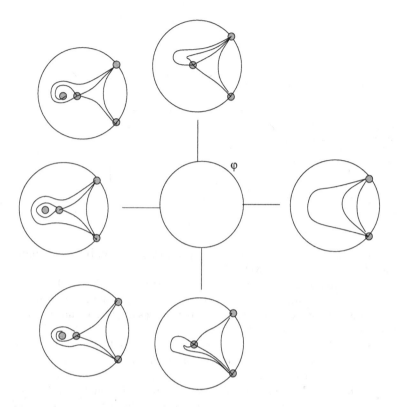

Figure 6.11

Here we can show easily that

$$\widetilde{G}^N = \alpha(\varphi)v^6 + O(r), \quad where \quad \alpha(\varphi) = -\frac{12}{5}\sqrt{6}\ \sin^2\varphi\ \ (\frac{9}{625}\ \sin^4\varphi + \cos\varphi).$$

Remark 36 *This constant $\alpha(\varphi)$ is a resonant term for the singular points at infinity of the quadratic vector field $X_\varphi = \overline{X}_{\varphi,\sigma} \mid r = 0$. It is clear that $\alpha(\varphi) = 0$ if $\varphi = 0, \pi$, when X_φ is a hamiltonian vector field. This value for α has been obtained directly by computation of the normal form using Maple. A cleverer way to obtain it would be to look at the complex extension of X_φ.*

From the normal form coordinates at p we deduce normal form coordinates for $\overline{X}_{\varphi,\sigma}$ in the "phase directional chart" ($\bar{x} = 1$), where we see the two points p_1, p_2. Let (r, v, Ω) be these coordinates near each point, with $r = v = \Omega = 0$ corresponding to p_1 and p_2 respectively.

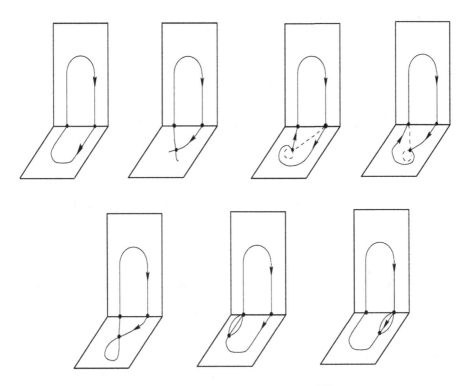

Figure 6.12 Limit periodic set of $\overline{X}$

As above, we take two sections Σ_1, Σ_2 near p_1, p_2 respectively, contained in $\{v = 1\}$ and belonging to the normal coordinate charts. We parametrize each of them by (Ω, u), with $u = r$ (the reason is that u is directly related to the initial parameters by $\mu = u^4 \cos \varphi$, $\nu = u \sin \varphi$). We can now give an asymptotic form for the transition map $R_{\varphi, \sigma}(\Omega, u)$:

Theorem 33 *Let γ be the Poincaré exponent for the cuspidal loop and $\alpha(\varphi)$ the resonant term, defined above. Then*

$$R_{\varphi, \sigma}(\Omega, u) = \gamma^{-1} \Omega + \sigma u^{-6} \alpha(\varphi)(\gamma^{-1} - 1)Lnu + O(uLn^2u), \qquad (6.10)$$

where the remainder $O(u Ln^2 u)$ is a function $\Phi(r, u, \sigma, \varphi)$ which is $\mathcal{C}^\infty$ in Ω and all of whose partial derivatives satisfy $\dfrac{\partial^k \Phi}{\partial \Omega^k} = O(uLn^2u)$ (we will say that the remainder term is $O(uLn^2u)$ in a "$\mathcal{C}^\infty$-sense").

Proof. I do not want to give a complete proof of this result, but just a rough outline. First, up to a $\mathcal{C}^1$-change of coordinates we can linearize the vector field at

p_1, p_2 (following a result of Belitskii; this is possible since there are no relations of the form $\lambda_i + \lambda_j = \lambda_k$ between the eigenvalues).

Taking transversal sections Σ'_1, Σ'_2 near p_1, p_2 in $\{r = 1\}$ belonging to the normal coordinate charts, we can write R as a composition of three maps,

$$R = T_2 \circ \overline{R} \circ T_1, \tag{6.11}$$

where T_1, T_2 are the transition maps near p_1 and p_2, and $\overline{R}$ is a $\mathcal{C}^1$ transition map from Σ'_1 to Σ'_2. We can write $\overline{R}_{\varphi,\sigma}(\Omega, u) = \sigma + \gamma^{-1}\,\Omega + O(u^2)$ and

$$T_1(\Omega) = u^6\,\Omega \ , \ T_2(\Omega) = u^{-6}\,\Omega = T_1^{-1}\,(\Omega). \tag{6.12}$$

Hence, R is conjugate to $\overline{R}$ by the linear map T_1, with coefficient $\left(\dfrac{u}{u_0}\right)^6 \to 0$ when $u \to 0$. This conjugacy has a *"funnelling effect"*: $R_{\varphi,\sigma}(\Omega, u)$ has to tend toward an affine map when $u \to 0$. We find

$$R_{\varphi,\sigma}\,(\Omega, u) = \sigma u^{-6} + \gamma^{-1}\,\Omega + O(u^6). \tag{6.13}$$

This argument, which was used in [R6], gives only a *"$\mathcal{C}^1$-control"* on the remainder $O(u^6)$. It would not be sufficient to study codimension 2 bifurcations and bifurcations of double cycles. To obtain a more accurate estimate we have to use the normal form given in Theorem 2. A first observation is that integration of such normal forms reduces to a 1-dimensional differential equation. Let $(r, \Omega) \in \Sigma \subset \{v = v_0\}$ and let $\varphi(t) = (\Omega(t), r(t), v(t))$ be the flow of $\overline{X}_{\varphi,\sigma}$ with these initial conditions. We have $r(t) = re^{-t}$, $v(t) = e^t$ and for $\Omega(t)$, the equation $\dot{\Omega} = 6\Omega + \widetilde{G}^N(\Omega, r, v, \lambda)$ (here $\lambda = (\sigma, \varphi)$). We look for $\Omega(t)$ in the form

$$\Omega(t) = e^{6t}\,\overline{\Omega}(t). \tag{6.14}$$

Substituting $r(t)$, $v(t)$ into (6.14), we obtain with $\Omega(0) = \Omega$,

$$\dot{\Omega}(t) = +6\Omega(t) + e^{+6t}\,\dot{\overline{\Omega}}(t) = 6\Omega(t) + \sum_{6i-j+\ell=6} \alpha_{ij\ell}\,e^{6t}\,\Omega(t)^i r^j, \tag{6.15}$$

and so

$$\dot{\overline{\Omega}}(t) = \sum_{6i-j+\ell} \alpha_{ij\ell}\,\overline{\Omega}^i\,r^j\,. \tag{6.16}$$

Therefore, $\overline{\Omega}$ is a solution of a 1-dimensional *autonomous equation*, $\dot{\overline{\Omega}}(t) = \widetilde{G}^N\,(\overline{\Omega}(t), r)$ (we no longer mention the dependence in $\lambda = (\varphi, \sigma)$).

Now, using classical estimates on the solution $\overline{\Omega}(t)$ we can prove that

$$\overline{\Omega}(t) = \Omega + \alpha\,t + O(rLn^2 r),$$

and then $T_1(\Omega) = u^6\,\overline{\Omega}\,(Ln\,u)$, with

$$\overline{\Omega}(Lnu) = \Omega + \alpha Lnu + C + O(uLn^2 u). \tag{6.17}$$

Here, C is a constant term and $O(u^2 Lnu)$ must be understood in the C^∞-sense. Up to this remainder we see that the unique change with respect to the previous formula (6.12) is the introduction of a translation term $\alpha Lnu + C$. Using a similar result for $T_2(\Omega)$, we obtain the asymptotic expression in the theorem for the composition $R = T_2 \circ \overline{R} \circ T_1$. $\qquad\square$

Remark 37 *Let $\tilde{\sigma} = \sigma \, u^{-6} \, \alpha(\varphi)(\gamma^{-1} - 1)Lnu$ be the rescaled translation term which enters into the expression of $R_{\varphi,\sigma}$. This term has to remain bounded. So, assume some $\tilde{\sigma}_0$ be chosen (depending on the choice of sections Σ_1, Σ_2) such that $\tilde{\sigma} \in [-\tilde{\sigma}_0, \tilde{\sigma}_0]$. We see that*

$$\sigma \in \alpha(\varphi)(\gamma^{-1} - 1)u^6 Lnu + u^6[-\tilde{\sigma}_0, \, \tilde{\sigma}_0].$$

This specifies a conic region in the parameter space (μ, ν, σ) of size u^6 around the cone $\sigma = \alpha(\varphi)(1-\gamma^{-1})u^6 Lnu$, in which are located all the bifurcations occurring in a neighborhood of Γ (recall that $\mu = u^4\cos\varphi$, $\nu = u \, \sin\varphi$). Of course, this remark does not refer to the small cycle or the Bogdanov-Takens theory near the saddle point.

We now study the function $G_{\varphi,\sigma} \, (r, u)$. If we put $g_\varphi(\Omega) = G_{\varphi,\sigma} \, (\Omega, 0)$ (independent of σ) in equation (6.9), we obtain

$$g_{\varphi,\sigma} \, (\Omega) = \gamma^{-1} \, \Omega + \tilde{\sigma}. \qquad (6.18)$$

As γ^{-1} is intrinsically defined, we also need to have an intrinsic definition of $\dfrac{\partial g}{\partial \Omega}$. Indeed, this follows from the fact that the variable Ω gives an *intrinsic metric structure* to the orbit space of X_φ. Let us study this point more closely. The vector field $X_\varphi(\dot{x} = y, \, \dot{y} = x^2 + \cos\varphi + y\sin\varphi)$ is extended to the Poincaré disk D by the blow-up formula $x = \dfrac{\cos\theta}{v^2}$, $y = \dfrac{\sin\theta}{v^3}$ which follows from the global blow-up. As we have noticed, the two singular points $p_1, p_2 \in \partial D^2$ have a resonant linear part with eigenvalues $\pm(1,6)$. Now, we have to define the map $g_\varphi(\Omega)$ as the transition map from Σ_1 to Σ_2. Recall that Σ_1, Σ_2 are intervals in $\{v = v_0\}$, for $u = 0$, and are parametrized by the variable Ω, which together with v is a normal form coordinate at p_1 or p_2. This means that at the point p_1, the vector field X_φ has the following normal form expression:

$$X_\varphi \begin{cases} \dot{v} &= v \\ \dot{\Omega} &= 6\Omega - \alpha v^6 \end{cases} \qquad (6.19)$$

The vector field has the opposite expression at p_2. Let us consider at p_1 two different reductions to the normal form H_1, H_2. Then, $H_1 \circ H_2^{-1}$ is a diffeomorphism which leaves invariant the equation (6.19). It is easy to prove that such a diffeomorphism can be written as

$$H_1 \circ H_2^{-1} \, (\Omega, v) = (\Omega + \beta v^6, v), \qquad (6.20)$$

for some $\beta \in \mathbb{R}$.

As a consequence, the variable Ω is uniquely defined (i.e., independent of the normalizing diffeomorphism) up to a translation.

Another way to say this is the following: let $\mathcal{O}_{p_1}$ be the space of trajectories with α-limit in p_1, in the half space $\{v > 0\}$. Then, the map $\Omega \to \gamma_\Omega$ (orbit in $\mathcal{O}_{p_1}$ through the point $(\Omega, 1)$ in the normal form charts) induces on the space $\mathcal{O}_{p_1}$ a well-defined metric (i.e., a parametrization defined up the choice of some origin). The same holds at the point p_2. Finally, we can see $g_\varphi(\Omega)$ as a map $g_\varphi(\Omega) : \mathcal{O}_{p_1} \to \mathcal{O}_{p_2}$ between the orbit-spaces at p_1 and p_2, defined by the transition. We call it the *"transition at ∞"* between p_1 and p_2. The spaces of orbits $\mathcal{O}_{p_1}, \mathcal{O}_{p_2}$ are isometric to $\mathbb{R}$, with variable Ω. Strictly speaking the map $g_\varphi(\Omega)$ is defined outside the interval of orbits which tend toward the possible attracting focus e_φ (when $\varphi \in [-\pi, -\frac{\pi}{2}])$, so let $D_\varphi \subset \mathbb{R}$ be the maximal domain of definition of $g_\varphi(\Omega)$,

$$D_\varphi = \mathbb{R} \text{ is } \varphi \in \left]-\frac{\pi}{2}, \frac{\pi}{2}\right[, \ D_\varphi = \mathbb{R} - \Omega_\varphi^1 \text{, if } \varphi \in \left[\frac{\pi}{2}, \pi\right],$$

(where Ω_φ^1 corresponds to the separatrix of s_φ) and $D_\varphi = \mathbb{R} -]\Omega_\varphi^1, \Omega_\varphi^2[$, where Ω_φ^1, Ω_φ^2 correspond to the two saddle separatrices of s_φ, if $\varphi \in] - \pi, -\frac{\pi}{2}]$ (see Figure 6.13).

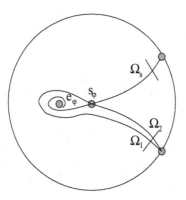

Figure 6.13

Finally, equation (6.9) for limit cycles has the following asymptotic form:

$$g_\varphi(\Omega) - \gamma^{-1} \, \Omega - \tilde{\sigma} + O(uLn^2 u) = 0,$$

(6.21)

$$\text{with } \tilde{\sigma} = \sigma \, u^{-6} + \alpha(\varphi)(\gamma^{-1} - 1)Lnu,$$

where the remainder $O(uLn^2 u)$ is $\mathcal{C}^\infty$ in Ω and is $O(u \, Ln^2 \, u)$ in a $\mathcal{C}^\infty$ sense. Hence, the bifurcations in the family depend only on the properties of $g_\varphi(\Omega)$.

6.2.3 Properties of $g_\varphi(\Omega)$

First, we will obtain a more tractable expression for $g_\varphi(\Omega)$. Observe that the interval of definition of any trajectory γ with $\alpha(\gamma) = p_1$ and $\omega(\gamma) = p_2$ is bounded, i.e., for any $m \in \gamma$, the trajectory tends to p_1, for $t \to \tau_1$, and to p_2, for $t \to \tau_2$, where τ_1, τ_2 are finite. We can call $\tau_2 - \tau_1 = T(\gamma)$ the transition time from p_1 to p_2 along γ.

Each orbit γ as above belongs to $D_\varphi \subset \mathcal{O}_{p_1}$, with its natural parametrization by Ω, so that we denote by $T_\varphi(\Omega)$ the transition time to go from p_1 to p_2 along the orbit of D_φ with parameter Ω. This function $T_\varphi(\Omega)$ is closely related to $g_\varphi(\Omega)$.

Proposition 16

$$\frac{\partial\, g_\varphi}{\partial\Omega}\, (\Omega) = \exp(\sin \varphi T_\varphi(\Omega)). \tag{6.22}$$

Proof. Let $\gamma_\varphi(\Omega)$ be the orbit of X_φ with parameter $\Omega \in D_\varphi$. Let Σ_1, Σ_2 be two sections in normal form charts near p_1, p_2 belonging to $\{v = v_0\}$. Let $G_\varphi^{v_0}(\Omega)$ denote the transition map between Σ_1 and Σ_2.

If we denote by $\| . \|$ the Euclidean norm in the normal form charts, it follows from a well known variational formula that

$$\frac{dG}{d\Omega}\, (\Omega) = \frac{\|X_\varphi(G_\varphi^{v_0}\, (\Omega))\|}{\|X_\varphi\, (\Omega)\|} \exp\left(\int_0^{T^{v_0}} \operatorname{div} X_\varphi dt\right), \tag{6.23}$$

where $\Omega \in \Sigma_1$, $G_\varphi^{v_0}(\Omega) \in \Sigma_2$ and T^{v_0} is the time necessary to go from Ω to $G_\varphi^{v_0}(\Omega)$. Now, $\operatorname{div} X_\varphi = \sin \varphi$ and $T^{v_0} = T_\varphi(\Omega) + O(v_0)$.

Moreover, it can be shown that

$$\frac{\|X_\varphi(G_\varphi^{v_0}\, (\Omega))\|}{\|X_\varphi\, (\Omega)\|} = 1 + O(v_0), \tag{6.24}$$

and that

$$\frac{\partial\, G_\varphi^{v_0}}{\partial\, \Omega}\, (\Omega) = \frac{\partial\, g_\varphi}{\partial\, \Omega}\, (\Omega) + O(v_0). \tag{6.25}$$

In order to prove the last estimate (6.25), we observe that $G_\varphi^{v_0}$ is conjugate to g_φ.

Finally, we have that

$$\frac{\partial\, g_\varphi}{\partial\, \Omega}(\Omega) = \exp(\sin \varphi T_\varphi(\Omega)) + O(v_0), \tag{6.26}$$

and the result follows because $\dfrac{\partial\, g_\varphi}{\partial\, \Omega}\, (\Omega)$ is in fact independent of v_0. $\qquad\square$

As the variable Ω is not explicitly defined, it is preferable to parametrize the transition time by the intersection point with the $0x$-axis. Let $M_\varphi \subset \mathbb{R}$ ($0x$-axis) be the set of points x such that α and ω-limit of the orbit through x are contained in $\{p_1, p_2\}$. We have a transition diffeomorphism $\Omega_\varphi(x)$ from M_φ to D_φ and $M_\varphi = \mathbb{R}$, if $\varphi \in]-\pi/2, \pi/2[$, $M_\varphi = \mathbb{R} - x_\varphi^1$ (x_φ^1: coordinate of the saddle s_φ), if $\varphi \in [\frac{\pi}{2}, \pi]$, and $M_\varphi = \mathbb{R}-]x_\varphi^2, x_\varphi^1[$ if $\varphi \in]-\pi, -\pi/2]$ where x_φ^1 is the coordinate of the intersection point of the left hand instable separatrix with $0x$, and x_φ^1 is the coordinate of s_φ (see Figure 6.14).

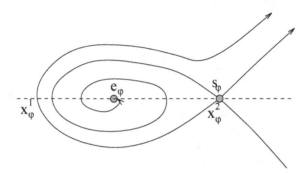

Figure 6.14

Definition 32 *We define the function time of transition* $t_\varphi(x) = T_\varphi(\Omega_\varphi(x))$, *for the orbit through* $x \in M_\varphi$.

We can now define the monotonicity property (M). First note that $t_\varphi(x) = t_{-\varphi}(g_{-\varphi})(x))$. This follows from the invariance of the differential equation of X_φ by $\varphi \to -\varphi$, $y \to -y$, $t \to -t$. Hence, it suffices to consider $\varphi \in [0, \pi]$

Property (M) for $t_\varphi(x)$.

1) For each φ, the function $x \to t_\varphi(x)$ has just one extremum (which is generic) at $x_\varphi \in \mathbb{R}$, when $\varphi \in [0, \pi/2[$; $t_\varphi(x)$ is monotonic, when $\varphi \in [\pi/2, \pi]$.

2) $\dfrac{d}{d\varphi}\, [t_\varphi(x_\varphi)] > 0$ for $\varphi \in \left[0, \dfrac{\pi}{2}\right[$.

Remark 38 *It is easy to verify that* $t_\varphi(x) \to 0$, *when* $x \to \pm\infty$. *It follows from this and from point 1) that* x_φ *is a maximum (in the case* $\varphi \in [0, \frac{\pi}{2}[$) *and that* $t_\varphi(x)$ *is increasing for* $x \in]-\infty, x_\varphi^1[$ *and* $t_\varphi(x)$ *is decreasing for* $x \in]x_\varphi^1, \infty[$ *(in the case* $\varphi \in \left[\frac{\pi}{2}, \pi\right]$). *Of course,* $t_\varphi(x) \to \infty$, *if* $x \to x_\varphi^1$ *or* x_φ^2.

As we have said above, this property (M) has not been proved yet, but is supported by strong numerical evidence. Moreover, it is possible to prove part 1) when $\varphi = 0, \pi$.

6.2.4 Monotonicity property for $t_0(x)$ and $t_\pi(x)$

When $\varphi = 0$, the vector field is a hamiltonian vector field with hamiltonian function
$$H(x,y) = \frac{1}{2} y^2 - \frac{x^3}{3} - x.$$
Up to an affine change in x, we can suppose that $H(x,y) = y^2 + x^3 + x$ and that the axis $0x$ is parametrized by the values h of the hamiltonian. Let Γ_h be the orbit $\{H = h\} \subset \mathbb{R}^2$. The transition time $t(h) = \displaystyle\int_{\Gamma_h} \frac{dx}{y}$.

Now, in the usual compactification of $\mathbb{C}^2 \subset \mathbb{P}_2(\mathbb{C})$, each Riemann surface $\{H = h\}$ has a regular point at infinity, so on this surface we can find a bounded cycle γ_n homotopic to Γ_h, and can make this choice continuously, for $h \in \mathbb{R}$. Then $\displaystyle\int_{\Gamma_h} \frac{dx}{y} = \int_{\gamma_n} \frac{dx}{y}.$

For such a continuous choice of $\gamma_h \subset \{H = h\}$, the abelian integral $J_0(h) = \int_{\gamma_h} \frac{dx}{y}$ has the properties described in 4.3.6. In particular, as a consequence of general formulas (4.63) in 4.3.6, the integrals I_0, I_1 or J_0, J_1 satisfy a Picard-Fuchs system. For instance, for J_0, J_1 we obtain

$$\begin{cases} \left(h^2 + \dfrac{4}{27}\right) J_0' = -\dfrac{1}{6} h J_0 - \dfrac{1}{9} J_1 \\ \left(h^2 + \dfrac{4}{27}\right) J_1' = -\dfrac{1}{27} J_0 + \dfrac{1}{6} h J_1. \end{cases} \tag{6.27}$$

We want to prove that $t'(h) = J_0'(h)$ has at most one simple root. To this end, we can use Petrov's lemma (Lemma 17 in 4.3.6). First, from the first line of (6.27) we see that $J_0' = 0$ if and only if $-\dfrac{1}{6} h J_0 - \dfrac{1}{9} J_1 = 0$ or equivalently if

$$g = \frac{J_1}{J_0} + \frac{3}{2} h = 0. \tag{6.28}$$

(Notice that J_0 has no root.)

From the Ricatti equation for $\dfrac{J_1}{J_0}$,

$$\left(h^2 + \frac{4}{27}\right)\left(\frac{J_1}{J_0}\right)' = \frac{1}{9} \left(\frac{J_1}{J_0}\right)^2 + \frac{1}{3} h\left(\frac{J_1}{J_0}\right) - \frac{1}{27} \tag{6.29}$$

deduced from (6.27), we can write a Ricatti equation for g,

$$\left(h^2 + \frac{4}{27}\right) g' = \frac{3}{4} g^2 + \frac{3}{4} h^2 + \frac{5}{27}. \tag{6.30}$$

The polynomial $R_0(h) = \frac{3}{4} h^2 + \frac{5}{27}$ has no real root. It thus follows from Lemma 17 in 4.3.6 that g has at most one simple root, and thus J_0' also has at most one simple root. Because $J_0 \to 0$ for $h \to \pm\infty$, it follows that J_0' has exactly one simple root where J_0 has a quadratic maximum.

In the case $\varphi = \pi$, we consider the hamiltonian $H(x, y) = y^2 - x + x^3$ introduced in Chapter 4. Now, the Ricatti equation for g has the form

$$\left(\frac{4}{27} - h^2\right) g' = R_2(h)g + R_1(h) + R_0(h),$$

(6.31)

$$\text{with } R_0(h) = \frac{1}{4} h^2 - \frac{1}{27}.$$

The roots of R_0 are precisely the singular values of $H : \left\{\pm\frac{2}{3\sqrt{3}}\right\}$. The orbits Γ_h, for $h \in \left[-\infty, +\frac{2}{3\sqrt{3}}\right[$, are on the left of the saddle point $s\left(x = -\frac{1}{3\sqrt{3}}\right)$, and surround the two critical points, for $h > \frac{2}{3\sqrt{3}}$ (see Figure 6.15).

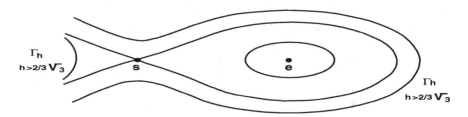

Figure 6.15

It is easy to verify that $t'(h) = J_0'(h) > 0$ for $h < \frac{2}{3\sqrt{3}}$. Now, J_0' has at most one simple root for $h \in \left]\frac{2}{3\sqrt{3}}, \infty\right]$, because R_0 has no roots for $h > \frac{2}{3\sqrt{3}}$. But, as $t(h) \to 0$ for $h \to \infty$ and $t(h) \to +\infty$ if $h \to \frac{2}{3\sqrt{3}}$ it follows that $t'(h)$ has also no root on $\left]\frac{2}{3\sqrt{3}}, \infty\right]$.

6.2.5 Indications for the proof of Theorem 31

We need to use the transition time $T_\varphi(\Omega)$ parametrized by $\Omega : T_\varphi(\Omega) = t_\varphi(x_\varphi(\Omega))$ where $t_\varphi(x)$ is the transition time parametrized by $x \in M_\varphi$ and $x_\varphi(\Omega)$ is the diffeomorphism $D_\varphi \to M_\varphi$. First, we remark that $T_\varphi(\Omega)$ has the same monotonicity property (M) as $t_\varphi(x)$:

1) $T_\varphi(\Omega)$ has just one extremum (a quadratic one) at $\omega_\varphi \in \mathbb{R}$ when $\varphi \in [0, \frac{\pi}{2}[$; $T_\varphi(\Omega)$ is decreasing when $\varphi \in [\frac{\pi}{2}, \pi]$.

2) $\frac{d}{d\varphi}\left[T_\varphi(\Omega_\varphi)\right] > 0$ when $\varphi \in [0, \frac{\pi}{2}[$.

To prove the first point, we have just to note that

$$\frac{\partial T_\varphi}{\partial \Omega} = \frac{\partial t_\varphi}{\partial x} \cdot \frac{\partial x_\varphi}{\partial \Omega} \text{ and that } \frac{\partial x_\varphi}{\partial \Omega} > 0.$$

Next, $T_\varphi(\Omega_\varphi) \equiv t_\varphi(x_\varphi)$ and point 2) is equivalent to the same property for t_φ.

The map $g_\varphi(\Omega)$ is defined up to a constant by the equation

$$\frac{\partial \Omega_\varphi}{\partial \Omega} = \exp(\sin \varphi \, T_\varphi(\Omega)).$$

The behavior which is illustrated in figure 6.16, where we represent the graphs $g_\varphi : \mathbb{R} \to \mathbb{R}$ for different values of φ, follows from property (M). The invariance of X_φ by the map $\varphi \to -\varphi$, $y \to -y$, $t \to -t$ implies that $g_\varphi^{-1} = g_{-\varphi}$ so that it suffices to consider $\varphi \in [0, \pi]$. In fact, $\frac{\partial g_\varphi}{\partial \Omega} < 1$ for $\varphi \in]-\pi, 0[$ and we have no bifurcation for these values of φ (see Figure 6.16).

Remark 39 *The value of the derivative of g_φ at the unique inflexion point of g_φ increases monotonically from 1 to ∞, when φ varies from 0 to $\frac{\pi}{2}$. This follows from*

$$\frac{\partial}{\partial \varphi}\left[\frac{\partial g_\varphi}{\partial \Omega}(\Omega_\varphi)\right] = \frac{\partial}{\partial \varphi} \exp(\sin \varphi t_\varphi(\Omega_\varphi))$$

$$= \left[\cos \varphi t_\varphi(\Omega_\varphi) + \sin \varphi \frac{\partial}{\partial \varphi}(t_\varphi(\Omega_\varphi))\right] \qquad (6.32)$$

$$\exp(\sin \varphi \, t_\varphi(\Omega_\varphi)),$$

so that $\frac{\partial}{\partial \varphi}\left[\frac{\partial g_\varphi}{\partial \Omega}(\Omega_\varphi)\right] > 0.$

Recall that the equation for the fixed points of the return map is equivalent to the shift equation

$$\Delta_{\varphi,\sigma}(\Omega, u) = g_\varphi(\Omega) - \gamma^{-1}\Omega - \tilde{\sigma} + O(uLn^2u) = 0.$$

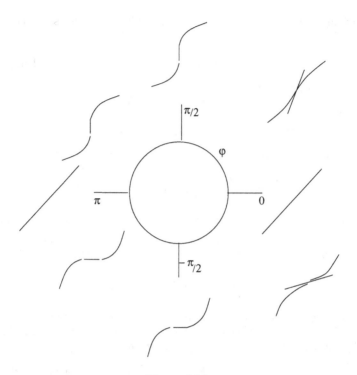

Figure 6.16

Graphically, we have to look at the intersection of the graph

$$g_{\varphi,\sigma}(\Omega, u) = g_\varphi(\Omega) + 0(uLn^2u)$$

with the line $R = \gamma^{-1}\,\Omega + \tilde{\sigma}$, where $\tilde{\sigma}$ is related to σ by $\tilde{\sigma} = u^{-6}\sigma + \alpha(\gamma^{-1}-1)Lnu$.

There exists a value $\varphi_1 \in \left]0, \dfrac{\pi}{2}\right[$ where the slope of the tangent at the inflexion point of g_{φ_1} is equal to γ^{-1}. The slope of the tangent at g_φ, for $\varphi < \varphi_1$ and any Ω is less than γ^{-1} and one has just one simple intersection point. There is a generic bifurcation of triple limit cycles for $\varphi = \varphi_1$.

For $u = 0$, this bifurcation is located at $(\varphi_1, \tilde{\sigma}_1)$, where $\tilde{\sigma}_1$ correspond to the intersection of the $\tilde{\sigma}$-axis with the tangent at the inflexion point of $g_{\varphi_1} : \tilde{\sigma}_1 = g_{\varphi_1}(\Omega_1) - \gamma^{-1}\,\Omega_{\varphi_1}$.

Hence, by the implicit function theorem, we have a line of bifurcation,

$$u \to (\tilde{\sigma}_1(u), \varphi_1(u))\ , \text{ with } \tilde{\sigma}_1(0) = \tilde{\sigma}_1,\ \varphi_1(0) = \varphi_1,$$

which defines a line TC in the 3-parameter space

$$u \to \left(\varphi_1(u),\ \sigma(u) = \alpha(\varphi_1(u))\ (\gamma^{-1}-1)u^6\ Ln\ u + u^6\ \tilde{\sigma}_1(u)\right),$$

or, explicitly,

$$(TC) \begin{cases} \mu(u) &= u^4 \cos(\varphi_1(u)) \\ \nu(u) &= u \sin(\varphi_1(u)) \\ \sigma(u) &= \alpha(\varphi_1(u))(\gamma^{-1} - 1)u^6 Ln\ u + u^6 \tilde{\sigma}_1(u). \end{cases} \tag{6.33}$$

From this line of triple cycles split two surfaces of double cycles. Their equation is given by

$$\frac{\partial g_\varphi}{\partial \Omega} = \gamma^{-1} + 0(u\ Ln^2\ u).$$

Or

$$\sin \varphi T_\varphi(\Omega) = Log\ \gamma^{-1} + 0(uLn^2u). \tag{6.34}$$

The curve $\Omega \to \sin \varphi T_\varphi(\Omega)$ has an unique extremum whose value increases with φ and tends to ∞, when $\varphi \to \dfrac{\pi}{2}$ (see Figure 6.17).

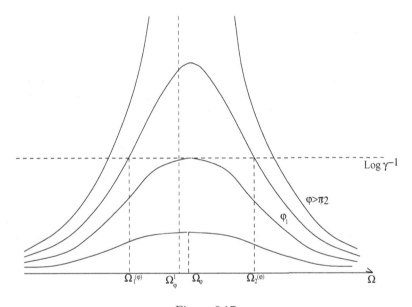

Figure 6.17

For $\varphi > \varphi_1$, the graph of g_φ cuts the horizontal $Log\ \gamma^{-1}$ in two points $\Omega_1(\varphi)$, $\Omega_2(\varphi)$ bifurcating from Ω_1, which define the two surfaces of double cycles which bifurcate from TC. These two surfaces may be extended for all values of $\varphi < \pi$.

When $\dfrac{\pi}{2} \le \varphi < \pi$ the graph g_φ has a discontinuity at Ω_φ^1 and a vertical gap $[g_\varphi^1, g_\varphi^2]$. The crossing of the line $\gamma^{-1}\Omega + \tilde{\sigma}$ at the point $(\Omega_\varphi^1, g_\varphi^1)$ corresponds to a saddle connection bifurcation L_s and the crossing at $(\Omega_\varphi^1, g_\varphi^2)$ to a saddle connection bifurcation L_r.

Hence, it is easy to justify all the bifurcations and the existence and the number of limit cycles, in a complement of a small neighborhood of $u = 0$, $\varphi = \pi$.

The study in such a neighborhood is in fact entirely independent of the hypothesis (M) and follows only from the local properties of $g_\varphi(\Omega)$ for φ near π.

For $\varphi = \pi$, we have $g_\pi(\Omega) \equiv \Omega$, but the convergence $g_\varphi(\Omega) \to \Omega$ is not regular when the value of Ω corresponds to a stable separatrix of the saddle s_π. To study the bifurcations at $\varphi = \pi$ we have to make a second directional blow-up in the parameter $\varphi = \pi - u\xi$ with $\xi \in [-\xi_0, \xi_0]$, $u \in \mathbb{R}^+$ small.

The vector field family $\overline{X}_{\varphi,u}$ is transformed to

$$\overline{X}_{\xi,u} \begin{cases} \dot{x} &= y \\ \dot{y} &= x^2 - 1 + uy(\xi + \varepsilon x) + 0(u^2). \end{cases} \tag{6.35}$$

This is a u-perturbation of the hamiltonian vector field X_0 with hamiltonian function $H(x,y) = \dfrac{1}{2}y^2 + x - \dfrac{x^2}{3}$, which we studied in Chapter 4 for the Bogdanov-Takens theory.

The map $\widetilde{G}_{\xi,\sigma}(\Omega, u) = G_{\pi - u\xi, \sigma}(\Omega, u)$ is a u-perturbation of the identity. It may be studied by composing regular maps with transitions near the saddle s on X_0. Before entering into the details of this computation, we draw the shape of the graph of $\widetilde{G}$ for $\widetilde{G}$, in Figure 6.18, for some small $u > 0$.

For ξ small, or large enough, we again find the shapes obtained for $g_\varphi(\Omega)$ for $\varphi > \pi$ and $\varphi < \pi$. These extremal shapes are the same for the two values of ε, but the deformation between these two shapes is not the same in the two cases. For instance, in the case $\varepsilon = 1$, the horizontal level in the central region of the graph of g_φ is surrounded by large slopes: this allows the possible existence of a pair of double cycles and also the possible existence of four limit cycles.

As $\widetilde{G} - \widetilde{G}_{|u=0}$ may be $O(u)$ (when Ω is a regular point) we need more precision on the remainder $0(uLn^2u)$ in $R_{\varphi,\sigma}(\Omega, u)$. In fact, we can prove that this term can be written

$$\psi_0(u) + \psi_1(u)\Omega + \psi_2(u, \Omega, \sigma, \varphi)\Omega^2, \tag{6.36}$$

with $\psi_0(u) = O(uLn^2u)$, $\psi_1(u) = O(uLn^2u)$ and $\psi_2 = O(u^5)$ $\quad$ (6.37)

(in the C^∞-sense).

Hence, we can write the transition map R,

$$R_{\varphi,\sigma}(\Omega, u) = (\gamma^{-1} + \psi_1)\Omega + (\sigma u^{-6} + \alpha(\gamma^{-1} - 1)Lnu + \psi_0) + O(u^5)\Omega^2. \tag{6.38}$$

Here the remainder is differentiable in Ω and can be incorporated in $\widetilde{G}$; we will write $\widetilde{R}_{\xi,\sigma}(\Omega, u)$ for R after the substitution $\varphi = \pi - u\xi$.

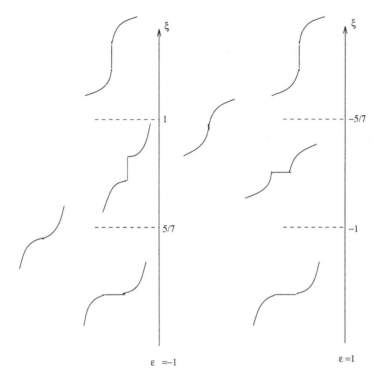

Figure 6.18

Bifurcations related to the double connection (DL) (see Figure 6.19).

As the separatrices of the saddle depend smoothly on the parameter, for (φ, u) near $(\pi, 0)$, we can choose the origins on Σ_1, Σ_2 in such a way that $\Omega = 0$ corresponds to the separatrix intersection. Let Ω_1 be the coordinate on Σ_2. We also introduce sections τ_1, τ_2, as in Figure 6.19, with coordinates z, z_1 respectively.

1) If $\Omega \geq 0$, $\widetilde{G}$ is the composition of the following three maps:

– From Σ_1 to τ_1,
$$z = \Omega^{(1-u\bar{\alpha})}(1 + \Phi_0), \qquad (6.39)$$
where $\bar{\alpha}(\xi, u, \ldots) = \xi + \varepsilon + 0(u)$ and Φ_0 is of class I (see Chapter 5, Section 5.1.3).

– From τ_1 to τ_2,
$$z_1 = u\bar{\beta}_0 + (1 + u\bar{\beta}_1)z[1 + \Phi_1], \qquad (6.40)$$
where $\bar{\beta}_0(\xi, u, \ldots) = a(\xi - \frac{5}{7}) + O(u)$.

Here, $u\bar{\beta}_0$ is the shift between the two "small separatrices" on the left of the saddle. This quantity was computed in Chapter 5 and $a = I_0\left(\frac{2}{3}\right)$ is the area of the singular hamiltonian disk. $\Phi_1 = O(u)$ is smooth.

– From τ_2 to Σ_2,

$$\Omega_1 = z_1^{1-u\bar{\alpha}}(1 + \Phi_2). \tag{6.41}$$

This is similar to the first transition.

By composition we obtain

$$\Omega_1 = \widetilde{G}_{\xi,\sigma}(\Omega, u) = (u\bar{\beta}_0 + (1 + u\bar{\beta}_1)\Omega^{1-u\bar{\alpha}}(1 + \widetilde{\Phi}_0))^{1-u\bar{\alpha}}(1 + \Phi_2) \tag{6.42}$$

for $\Omega \leq 0$.

The functions $\widetilde{\Phi}_0$, Φ_2 are of class I and are equal to zero for $u = 0$.

2) If $\Omega \leq 0$, we only have the transition near the saddle

$$\Omega_1 = \widetilde{G}_{\xi,\sigma}(\Omega, u) = \Omega^{(1-u\bar{\alpha})}(1 + \Phi_3). \tag{6.43}$$

The equation for the connection L_ℓ (small connection) is given by $\bar{\beta}_0 = 0$, i.e.,

$$\xi = \xi_0(u, \sigma) = \frac{5}{7} + O(u). \tag{6.44}$$

The equation for the right-hand connection L_r is given by writing $\widetilde{R}_{\xi,\sigma}(0, u) = 0$. This gives

$$\sigma = \sigma_0(u, \xi) = \alpha(\pi - u\xi)u^6 Lnu(1 + O(u)). \tag{6.45}$$

The two equations (6.44) and (6.45) define two transversal surfaces. They intersect along the line DL (double loop bifurcation).

To study the bifurcations near DL it is useful to introduce a local parameter $(\bar{\xi}, \bar{\sigma})$ by

$$\xi = \bar{\xi} + \xi_0, \quad \sigma = \bar{\sigma} + \sigma_0. \tag{6.46}$$

In this local parametrization, L_ℓ and L_r correspond to $\bar{\xi} = 0$ and $\bar{\sigma} = 0$ respectively. It is now easy to compute equations for the other surfaces of bifurcation. For instance, the "lower connection" L_i which corresponds to $\widetilde{R} = \widetilde{G} = 0$ verifies the equation

$$(L_i) \; : \; \bar{\sigma} = \gamma^{-1}au^{7-\bar{\alpha}_1 u}\, \bar{\xi}^{\frac{1}{1-\bar{\alpha}_1 u}}\,(1 + O(u)). \tag{6.47}$$

We also have

$$(L_s) \; : \; \bar{\sigma} = au^{7-\bar{\alpha}_1 u}\, \bar{\xi}^{1-\bar{\alpha}_1 u}\,(1 + O(u)) \tag{6.48}$$

for L_s (the "upper connection") given by $\Omega = 0$, $\widetilde{G} = R$.

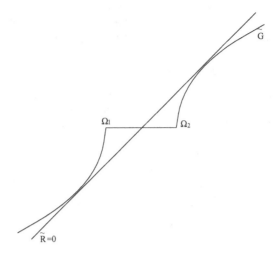

Figure 6.19

In the case $\varepsilon = 1$, we can look for a pair of double limit cycles in the local chart $(\bar{\xi}, \bar{\sigma})$. Here, we only present a heuristic computation (see [DRS3] for a rigorous treatment).

We write that the "line" $\widetilde{R} = 0$ has a double tangency with the $\widetilde{G}$-graph (see Figure 6.20).

The length of the horizontal level $[\Omega_1, \Omega_2]$ is of order $u \mid \beta_0 \mid^{\frac{1}{1-u\bar{\alpha}_1}}$. In a first approximation the graph of $\widetilde{G}$ is symmetrical, and one has a double tangent when this tangent passes through the middle of $[\Omega_1, \Omega_2]$. Putting this into the equation of $\widetilde{G}$, we obtain

$$| \bar{\xi} |= C_3(1 + O(u))e^{-\frac{\mathrm{Log}(\gamma^{-1})}{u}}, \tag{6.49}$$

for $C_3 = 2\gamma^{-1}a^{-1}$.

This is the $\bar{\xi}$-coordinate of the crossing point of the lines $DC^{in}(u)$ and $DC^{out}(u)$, for a fixed value of u.

To obtain the $\bar{\sigma}$-coordinate of $DC^{in}(u) \bigcap DC^{out}(u)$, we have to compute the intersection of the double tangent and the $\bar{\sigma}$-axis.

We find

$$\bar{\sigma} = -\gamma^{-1}(1 + O(u))u^7 e^{-\frac{\mathrm{Log}(\gamma^{-1})}{u}}. \tag{6.50}$$

A more striking fact is that the two lines DL and $DC^{in} \bigcap DC^{out}$ have a flat contact of order $e^{-\frac{\mathrm{Log}(\gamma^{-1})}{u}}$. Therefore, the rectangular region in Figure 6.5, corresponding to the existence of four large limit cycles, is flat in u (the "radius" of the intersection sphere in the parameter space).

Remark 40 *The above considerations are only heuristic. A rigorous treatment presented in [DRS3] leads to a different value for the coefficient in the exponential in the formulas (6.49), (6.50):* $\frac{7}{24} Log(\gamma^{-}1)$ *in place of* $Log(\gamma^{-}1)$.

Saddle connection of codimension 2

Let us consider, for instance, the case $\varepsilon = 1$. The codimension 2 saddle connections correspond to Trace (saddle)=0. They correspond to lines starting from $u = 0$, $\xi = -1$ (the hyperbolicity ratio is equal to $1 - u\bar{\alpha}_1$ with $\bar{\alpha}_1 = 1 + \xi + O(u)$. We look, for instance, for connections of type L_s. Notice that $\Omega = 0$ corresponds to the entering separatrix. We can write

$$\widetilde{G} = \Omega + u(\bar{\alpha}_0(\xi) + \bar{\alpha}_1(\xi)[\Omega\omega + \cdots] + \bar{\alpha}_2(\xi)\Omega + \cdots) + O(u^2), \qquad (6.51)$$

with the conventions introduced in Chapter 5. We know that $\bar{\alpha}_0(1) > 0$.

The equation for limit cycles $\widetilde{G} = \widetilde{R}$ gives

$$(h - 1 + \psi_1)\Omega + (\sigma u^{-6} - \alpha(\gamma^{-1} - 1)Lnu + \psi_0)$$

$$-u(\bar{\alpha}_0 + \bar{\alpha}_1[\Omega\omega + \cdots] + \bar{\alpha}_2\Omega) + O(u^2)O(\Omega) = 0. \qquad (6.52)$$

We can rearrange this equation into

$$\widetilde{\alpha}_0 + \widetilde{\alpha}_1[\Omega\omega + \cdots] + \widetilde{\alpha}_2\Omega + O(\Omega) = 0, \qquad (6.53)$$

with

$$\widetilde{\alpha}_0 = \sigma u^{-6} - \alpha(\gamma^{-1} - 1)Lnu + O(uLn^2u) \qquad (6.54)$$

$$\widetilde{\alpha}_1 = -u\bar{\alpha}_1 + O(u^2) \qquad (6.55)$$

$$\widetilde{\alpha}_2 = \gamma^{-1} - 1 + O(u). \qquad (6.56)$$

We can see that the bifurcation is generic for $u \neq 0$ ($\widetilde{\alpha}_2 \neq 0$, $\widetilde{\alpha}_1$ and $\widetilde{\alpha}_0$ are independent functions of σ and ξ). To verify this property, it suffices to express the parameters σ, ξ as functions of the versal ones $\widetilde{\alpha}_0$, $\widetilde{\alpha}_1$:

$$\begin{cases} \sigma & = \alpha(\gamma^{-1} - 1)uLn^2u + u^6\widetilde{\alpha}_0 + O(u^7Ln^2u) \\ \xi - 1 & = -u^{-1}\widetilde{\alpha}_1 + O(u). \end{cases} \qquad (6.57)$$

A fixed rectangle in the $(\widetilde{\alpha}_0, \widetilde{\alpha}_1)$-plane corresponds to a rectangle in the (ξ, σ)-plane with a σ-dimension of order u^6 and the ξ one of order u^{-1}; this domain degenerates when $u \to 0$.

For a codimension 2 connection of type L_i, the corresponding formulas are

$$\begin{cases} \widetilde{\alpha}_0 & = u^{-6}\sigma - \alpha(\gamma^{-1} - 1)Lnu + O(uLn^2u) \\ \widetilde{\alpha}_1 & = -2u\bar{\alpha}_1 + O(u^2) \\ \widetilde{\alpha}_2 & = \gamma^{-1} - 1 + O(u) \end{cases} \qquad (6.58)$$

Remark 41 *Formulas (6.58) for the L_i-connection are almost identical to those for the L_s-connection, up to the remainder. They do not permit us to distinguish between them. This corresponds to the fact that the coefficient $\bar{a}_0(1)$ whose sign distinguishes the two cases, is absorbed in the remainder. To obtain an estimate of the separation, a more precise computation is needed.*

The regular values of (ξ, Ω)

If we take a value (ξ, Ω) where Ω is a regular value of $\widetilde{G}_{\xi,\sigma}(\Omega, u)$ at $u = 0$, then the function $\widetilde{G}$ tends $\mathcal{C}^1$ to the identity, for $u \to 0$. As $\widetilde{R}$ converges to $\gamma^{-1}\Omega + \widetilde{\sigma}$, with $\gamma^{-1} \neq 1$, we can only have simple roots in a compact domain which does not contain singular values of Ω: all the bifurcations at $\varphi = \pi$ happen for $\xi = -\varepsilon$, $-\frac{5}{7}\varepsilon$ and for the Ω-value of the entering separatrix. They are precisely the cases we have considered above.

This finishes the outline of the proof of Theorem 31. A more complete study is given in [DRS3].

6.3   A method of desingularization for analytic vector fields

In this section, we consider an analytic family (X_λ) on S^2. We explain how the generalized blowing-up may be used to give a general method of desingularization for the families (X_λ), as presented in [DeR]. The geometrical object we obtain by such desingularization is called a *foliated local vector field*. Roughly speaking, it is given by a *local vector field* (another name for dimension 1 foliations with singularities). This local vector field is tangent to a singular 2-dimensional foliation, which comes from the blowing-up of the fibration of $\mathbb{R}^{k+2}$ onto the parameter space $\mathbb{R}^k$.

Next, we define *desingularization operations* precisely. Apart from a generalized blow-up (which generalizes the global blow-up used in the previous sections), we introduce the possibility *of dividing* the local vector field by local functions (for instance the function r in the last paragraph) and also to replace a family by a new one *inducing* the first one. Finally, we propose general conjectures concerning desingularizations of analytic families. We present the relationship of our theory with Trifonov's theory and assess the chances of success in proving the conjectures.

In what follows all objects we consider, such as families, maps and so on, will be *real analytic*.

6.3.1   Foliated local vector fields

Definition 33 *Let E be a compact analytic manifold, with a possible non-empty boundary. A* local vector field *is defined on E by a finite open covering $\{U_i\}$ with some analytic vector field X_i on each U_i, verifying the following compatibility con-*

dition: for each pair of indices i, j such that

$$U_i \cap U_j \neq \emptyset,$$

there exists an analytic function g_{ij} defined and strictly positive in $U_i \cap U_j$ such that

$$X_i = g_{ij} X_j \quad on \ U_i \cap U_j.$$

Two collections $\{U_i, X_i\}$ and $\{V_j, Y_i\}$ as above are said to be equivalent if there exist positive analytic functions f_{ij} defined on $U_i \cap V_j$ such that $X_i = f_{ij} Y_j$. A *local vector field* on E is an equivalence class. Of course, each vector field defines a local vector field.

We denote by $Z(X)$ the union of sets of singular points of all X_i associated with a local vector field X. This set does not depend on the choice of the collection $\{U_i, X_i\}$.

Remark 42 *Such a local vector field is often called an oriented singular 1-dimensional foliation. We prefer the terminology "local vector field" because we reserve the term "foliation" for another purpose (see below).*

Definition 34 *We will call* singular fibration *a triple (E, π, Λ) consisting of*

- *a compact real analytic manifold E of dimension $k + 2$,*
- *a compact real analytic manifold Λ of dimension k,*
- *an analytic surjective mapping $\pi : E \rightarrow \Lambda$ such that for each $x \in E$ there are local coordinates $x_1, x_2, \ldots, x_{k+2}$ in a neighborhood of $\pi(x)$, sending $\pi(x)$ to 0, where π takes the form*

$$\lambda_1 = \prod_{i=1}^{k+2} x_i^{p_1^i}, \ldots, \lambda_k = \prod_{i=1}^{k+2} x_i^{p_k^i} \quad with \ p_j^i \in \mathbb{N}. \tag{6.59}$$

We suppose moreover, that π is regular (rank $(\pi) = k$) on an open dense set U_0. We suppose also that each regular fiber of π in U_0 is diffeomorphic to a 2-dimensional compact submanifold of the 2-sphere (possibly with non-empty boundary), i.e., a surface of genus 0.

Observe that for a foliation defined on an open dense set in E, any two extensions coincide on the intersection of their domains. Then there exists a unique maximal foliation extending the given one. We apply this remark here and we will denote by $\mathcal{F}$ the *maximal foliation* extending the foliation $\mathcal{F}_0$ defined by the connected components of the regular fibers of π on U_0. The domain of $\mathcal{F}$ will be denoted by U and the singular set of $\mathcal{F}$, $E \backslash U$ will be denoted by Σ.

Proposition 17 [DeR] *The maximal foliation $\mathcal{F}$, associated with a singular fibration (E, π, Λ) verifies the following properties:*

1. *The set Σ is an analytic subset of E and it is decomposed into two analytic manifolds Σ_1 and Σ_0, with cod $\Sigma_1 \geq 1$ and cod $\Sigma_0 \geq 2$, such that $\partial\Sigma_1 = \Sigma_0$.*
2. *If L is a leaf of $\mathcal{F}$, $\overline{L}$ is homeomorphic to a closed submanifold of S^2, with boundary and corners. For a given leaf L, let $\partial_0 L$ stand for the set of corners and $\partial_1 L = \partial L \backslash \partial_0 L$.*
3. *We have $U_L \partial L = \Sigma$, and $\{\partial_1 L, L \in \mathcal{F}\}$ defines an analytic foliation of Σ_1. Also, $U \partial_0 L = \Sigma_0$, which can be foliated by points.*
4. *Let $x \in \Sigma$ and let $\mathcal{L}_x$ be the collection of all the leaves L of $\mathcal{F}$ such that $\overline{L}$ is an analytic manifold with boundary in some neighborhoods of x. Suppose that $x \in \Sigma_1$ and let ℓ_x denote the leaf of Σ_1 through x. We have*

$$\bigcap\{T_x L \mid L \in \mathcal{L}_x \text{ and } x \in \overline{L}\} = T_x \ell_x. \tag{6.60}$$

Similarly, for $x \in \Sigma_0$, we have

$$\bigcap\{T_x L \mid L \in \mathcal{L}_x \text{ and } x \in \overline{L}\} = \{0\}. \tag{6.61}$$

Definition 35 *Given a singular fibration (E, π, Λ) and a point $x \in E$, a leaf through x is*

- *the 2-dimensional leaf of $\mathcal{F}$ containing x, if $x \in U$*
- *the 1-dimensional leaf of Σ_1 containing x, if $x \in \Sigma_1$*
- *the point $\{x\}$, if $x \in \Sigma_0$.*

We will introduce the main object of this paragraph, generalizing the notion of a family of vector fields.

Definition 36 *A foliated local vector field $\mathcal{E} = (E, \pi, \Lambda, X)$ is an object consisting of a singular fibration (E, π, Λ) and a local vector field defined on E such that X is tangent to the fibers, i.e., $d\pi(x)[X(x)] = 0$ for all $x \in E$.*

Remark 43 *An example of a foliated local vector field is given by any analytic family X_λ on S^2, with λ belonging to some compact analytic manifold Λ; $E = S^2 \times \Lambda$ and π is the natural projection of E on Λ and X is the family seen as a vector field on E.*

Proposition 18 *A foliated local vector field X is tangent to the leaf through x, at each point $x \in E$.*

Proof. The vector field X is tangent to the regular fibers by definition, so by continuity and by the density of U, it is tangent to all the leaves of the maximal foliation $\mathcal{F}$.

Now take a point $x \in \Sigma_1$. By continuity and using point 4 of Proposition 17, we get that $X(x) \in T_x \ell_x$ where ℓ_x is the leaf through x. The same holds when $x \in \Sigma_0$. $\qquad\square$

Of course, the notion of a foliated local vector field is motivated by global blow-up as used in Section 6.2 above. For instance, let $E = S^2 \times S^2$ and let $\mathbb{R}^2 \times \mathbb{R}^2$ be a local chart at some point $(x_0, \lambda_0) \in S^2 \times S^2$ (with $x_0 = 0$, $\lambda_0 = 0$). The usual blowing-up at this point is a map $\Phi : \widehat{E} \to E$. The new space $\widehat{E}$ has a singular projection $\widehat{\pi} = \pi \circ \Phi$ on $\Lambda \simeq S^2$. The critical locus of Φ is a 3-sphere D. For regular values of $\widehat{\pi}$, $\lambda \in S^2 - \{\lambda_0\}$, we have regular fibers $\widehat{\pi}^{-1}(\lambda) \simeq S^2$; on the other hand $\widehat{\pi}^{-1}(\lambda_0) = D \cup \widehat{F}$ (where $\widehat{F}$ is the blowing-up of the fiber $F = \pi^{-1}(\lambda_0)$. The foliation $\mathcal{F}_0$, whose leaves are the fibers $\widehat{\pi}^{-1}(\lambda)$ for $\lambda \neq \lambda_0$, extends to a maximal foliation $\mathcal{F}$. The set $\Sigma \simeq S^1$ is the common boundary of fibers $\widehat{F} \simeq D^2$. This set Σ is contained in D and $D \backslash \Sigma$ is made of 2-disks foliating $D \backslash \Sigma$ (see Figure 6.21).

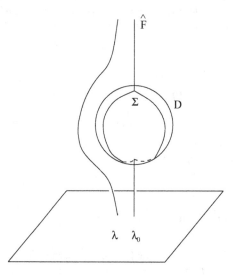

Figure 6.20

6.3.2   Operations of desingularization

A *limit cycle* of a foliated local vector field is a limit cycle of the restriction X_L of X to some leaf of $\mathcal{F}$.

In this section, we define operations to pass from one foliated local vector field to another, keeping control of the number of limit cycles we have in each leaf.

Definition 37 *We say that K is a* uniform bound *on the number of limit cycles for a foliated local vector field $\mathcal{E}$ (in short a* uniform bound *for $\mathcal{E}$) if, for each L belonging to the maximal foliation $\mathcal{F}$, the number of limit cycles of the restricted field X_L is not greater than K. If such a bound exists for a given foliated local vector field $\mathcal{E}$, we say that $\mathcal{E}$ has the* finiteness property.

The most important property of the three operations described below is that *they preserve the finiteness property*, i.e., if a uniform bound exists for $\widetilde{\mathcal{E}}$, obtained from $\mathcal{E}$ by one of the desingularization operation, then $\mathcal{E}$ also has the finiteness property.

Induction

Definition 38 *We say that $\mathcal{E} = (F \times \Lambda, \pi, \Lambda, X_L)$ is induced by another family $\widetilde{\mathcal{E}} = (F \times \widetilde{\Lambda}, \widetilde{\pi}, \widetilde{\Lambda}, X_{\widetilde{\lambda}})$ if there exists an analytic map $\Phi : \Lambda \to \widetilde{\Lambda}$ such that, for any λ, (X_λ) is topologically equivalent to $(\widetilde{X}_{\Phi(\lambda)})$.*

Observe that if $\widetilde{K}$ is a bound for $(\widetilde{X}_{\widetilde{\lambda}})$, then it is also a bound for (X_λ).

Local division

Definition 39 *Given two local vector fields X, Y on E, we say that Y is the* result *of a local division of X if there is a finite open covering $\{V_i\}$ of E, for which both X and Y are defined by local vector fields X_i, Y_i and analytic functions $f_i : V_i \to \mathbb{R}$ such that $X_i = f_i Y_i$ on V_i.*

The functions f_i may have zeros and Y may have more limit cycles than X_i, but any limit cycle of X is also a limit cycle for Y, so that if Y has the finiteness property, so does X.

Generalized blow-up

First, we define the global blow-up at a point in the context of foliated local vector field. Let $\mathcal{E} = (\varepsilon, \pi, \Lambda, X)$ and $p \in E$. Suppose we choose local coordinates $(x_1, \ldots, x_n)$ around p, and $(\lambda_1, \ldots, \lambda_k)$ around $\pi(p)$ where π is a monomial projection (as above in Definition 34). Suppose a system of weights $(\alpha_1, \ldots, \alpha_n) \in \mathbb{N}^n$ is chosen. We can define a map in coordinates

$$\varphi : S^{n-1} \times \mathbb{R}^+ \to \mathbb{R}^n , \quad \text{by} \quad x = \tau^\alpha \, \bar{x} = (\tau^{\alpha_1} \, \bar{x}_1, \ldots, \tau^{\alpha_n} \, \bar{x}_n).$$

We define the blown-up space

$$\widetilde{E} = (E \backslash \{p\}) \bigcup S^{n-1} \times \mathbb{R}^+ / \sim,$$

where $\sim$ is the identification $(\bar{x}, \tau) \in S^{n-1} \times \mathbb{R}^+ \sim T \circ \varphi(\bar{x}, \tau) \in E$ and T is the coordinate map $T : \mathbb{R}^n \to E$.

Now the blow-up map Φ is defined by the commutative diagram

$$
\begin{array}{ccc}
S^{n-1} \times \mathbb{R}^+ & \xrightarrow{\varphi} & \mathbb{R}^n \\
{\scriptstyle i} \downarrow & & \downarrow {\scriptstyle T} \\
\widetilde{E} & \xrightarrow{\Phi} & \mathbb{R}^n
\end{array}
\qquad (6.62)
$$

We denote by D the set $i(S^{n-1} \times \{0\}) \subset \widetilde{E}$ which is the *critical locus* of Φ.

We consider the effect of such a blow-up on a vector field defined near p. So, let X be a vector field in $W = T(\mathbb{R}^n)$ with $X(p) = 0$. Then, an easy computation shows that there exists $s \in \mathbb{Z}$ such that $\tau^s \, \Phi_*^{-1}(X) = \overline{X}$ is an analytic vector field on $S^{n-1} \times \mathbb{R}^+$. We take the *minimal such* s. If $\mathcal{E} = (E, \pi, \Lambda, X)$ and $X(p) = 0$, the above operation gives a well-defined local vector field $\widetilde{X}$ on $\widetilde{E}$, such that $\widetilde{X}_{|\widetilde{E}\backslash D} \simeq X_{|E\backslash\{p\}}$. Finally, the *blown-up foliated local vector field of* $\mathcal{E}$ is equal to $\widetilde{\mathcal{E}} = (\widetilde{E}, \widetilde{\pi}, \widetilde{\Lambda}, \widetilde{X})$, where $\widetilde{\Lambda} = \Lambda$ and $\widetilde{\pi} = \pi \circ \Phi$.

One can generalize the blow-up operation by replacing p by some compact submanifold $C \subset E$. If cod $C = n$, this submanifold is subject to some restrictions: in order to blow-up with a given system of weights $(\alpha_1, \ldots, \alpha_n)$, the embedding $C \subset E$ must have the so-called *admissible trivialization* by an atlas of charts $W_i \simeq U_i \times \mathbb{R}^n$, where U_i is a chart in C and the projection π on each W_i is monomial in the normal factor $\mathbb{R}^n$. See definitions and details in [DeR].

If the above conditions are fulfilled and if C is contained in $Z(X)$, the set of zeros of X, then we can blow-up E along C, using the weights $(\alpha_1, \ldots, \alpha_n)$. We produce a new foliated local vector field $\widetilde{\mathcal{E}} = (\widetilde{E}, \widetilde{\pi}, \Lambda, \widetilde{X})$ where

- $\widetilde{E}$ is the blown-up space,
- $\widetilde{\pi} = \pi \circ \Phi$, where $\Phi : \widetilde{E} \to E$ is the blowing-up map,
- $\widetilde{X}$ is a local vector field on $\widetilde{E}$ which is equal to $\Phi_*^{-1}(X)$ on $\widetilde{E}\backslash D$, where D is the critical locus of Φ (D is a fibered bundle over C, with fiber diffeomorphic to S^{n-1}).

As $C \subset Z(X)$, we do not destroy any limit cycles of X. The blow-up may disconnect some leaves, but each leaf of $\mathcal{E}$ is covered by a uniform finite number of leaves of $\widetilde{\mathcal{E}}$, so that the finiteness property for $\widetilde{\mathcal{E}}$ implies the finiteness property for $\mathcal{E}$.

6.3.3 Conjectures

Definition 40 *Let* $\mathcal{E} = (E, \pi, \Lambda, X)$ *be a foliated local vector field and* $p \in E$ *be a singular point of* X *(*$p \in Z(X)$*). We say that* p *is* an elementary singular point *if it is an elementary singular point of* X_L, *for each* $L \in \mathcal{L}_p$, *(i.e., the 1-jet* $J^1 X_L(p)$ *has at least a non-zero real eigenvalue). Note that* p *may belong to the critical set* Σ, *and in this case* $p \in \overline{L}$ *means that* $p \in \partial L$. $\mathcal{L}_p$ *may contain many leaves* L.

We say that a compact invariant set Γ *is a* limit periodic set *of* $\mathcal{E}$ *if there exists a sequence of limit cycles of* $\mathcal{E}$ *converging to* Γ *in the sense of the Hausdorff topology of the topological space* $\mathcal{C}(E)$ *of compact, non-empty subsets in* E. *A limit periodic set is said to be* elementary *if each of its points is either regular* ($X(p) \neq 0$) *or is an elementary singular point.*

Remark 44 *It seems possible to show, using Poincaré-Bendixson theorem as in Chapter 2, that each elementary limit set* Γ *is formed by a finite number of arcs,* *each of them being either a regular trajectory or a normally hyperbolic arc of zeros, and a finite number of isolated singular points. Moreover, each of these arcs would be contained in the closure of one of the leaves of $\mathcal{F}$, and some singular points in Γ could be in Σ.*

Such a curve is similar to the graphics of families introduced in Chapter 2, with the difference that Γ can now pass through several leaves and the critical set Σ.

Such elementary limit sets may be see in Figure 6.12 above.

Definition 41 *A desingularization step is a correspondence*

$$\{\mathcal{E}_i\}_{i \in I} \to \{\mathcal{E}_{ij}\}_{(i,j) \in I \times J}$$

between two collections of foliated local vector fields, satisfying the following conditions. Let $\mathcal{E}_i = (E_i, \pi_i, \Lambda_i, X_i)$. We suppose that there exists a collection

$$\{\mathcal{E}_{ij}^\circ\}_{(i,j) \in I \times J}, \quad \mathcal{E}_{ij}^\circ = (E_{ij}, \pi_{ij}, \Lambda_{ij}, X_{ij})$$

such that

(1) *$E_{ij} \subset E_i$ for all $(i,j) \in I \times J$,*

(2) *for each $i \in I$, every non-elementary limit periodic set of $\mathcal{E}_i$ is contained in the interior of one of E_{ij},*

(3) *for each $(i,j) \in I \times J$, the maximal foliation of $\mathcal{E}_{ij}^\circ$ is the trace on E_{ij} of the maximal foliation of $\mathcal{E}_i$. Moreover, there is an analytic map $\varphi_{ij} : \Lambda_{ij} \to \Lambda_i$ such that $\pi_i = \varphi_{ij} \circ \pi_{ij}$ in restriction to E_{ij},*

(4) *for each $(i,j) \in I \times J$, $\mathcal{E}_{ij}$ is either equal to $\mathcal{E}_{ij}^\circ$ or is induced from $\mathcal{E}_{ij}^\circ$ by one of the three desingularization operations of the previous section.*

Using the above conditions, we can now formulate two conjectures:

Desingularization Conjecture

For any analytic family (X_λ) on S^2, with λ belonging to some compact analytic manifold Λ, there exists a finite number of desingularization steps such that in the resulting final collection $\{\mathcal{E}_i\}$ of foliated local vector fields any limit periodic set is elementary.

Reduced Local Finite Cyclicity Conjecture

Any elementary limit periodic set Γ of an analytic foliated local vector field $\mathcal{E} = (E, \pi, \Lambda, X)$ has the finite cyclicity property, i.e., there exists $\varepsilon > 0$ and $K \in N$ such that the number of limit cycles of $\mathcal{E}$, at E-Hausdorff distance to Γ less than ε, is bounded by K for each leaf $L \in \mathcal{F}$. We suppose here that some metric is chosen on E.

Now, as for the compact families in Chapter 2, the finite cyclicity property for each limit periodic set implies the finiteness property for $\mathcal{E}$:

Lemma 24 *([DeR]). If each limit periodic set of a foliated local vector field $\mathcal{E}$ has finite cyclicity, then $\mathcal{E}$ has the finiteness property.*

A finiteness conjecture for compact analytic families was formulated in Chapter 2 (Section 2.2). We can now replace it by a more elaborate one, in fact by the above two conjectures.

Proposition 19 *The Desingularization Conjecture together with the Reduced Local Finite Cyclicity Conjecture implies that each compact analytic family on $S^2 \times \Lambda$ has the finiteness property.*

Proof. Suppose that after k desingularization steps we have obtained a final collection of foliated local vector fields whose limit periodic sets are elementary. It follows from the above lemma and the second conjecture that each foliated local vector field $\mathcal{E}$ in this collection has the finiteness property.

To finish the proof, we repeat this argument inductively. Suppose that we have proved that in the s-th step of desingularization, $s \geq 1$, all foliated local vector fields have the finiteness property. Let $\mathcal{E}$ be one of the foliated local vector field of the $(s-1)$-th step. Since the finiteness property is preserved by the three desingularization operations, each non-elementary limit periodic set of $\mathcal{E}$ has the finite cyclicity property. Each elementary limit periodic set has it too, as follows from the second conjecture. So, $\mathcal{E}$ has the finiteness property as a consequence of Lemma 24. Therefore, by induction, we obtain the required result. $\square$

6.3.4   Final comments and perspectives

The desingularization method was useful in the study of several unfoldings and in obtaining their diagrams of bifurcation. In Section 6.2 we presented a detailed review of the study of generic unfoldings of cuspidal loops, to appear in the forthcoming publication [DRS3]. Some other examples of the desingularization method were given in [DR2] for the unfolding of codimension 3 nilpotent focus and for Van der Pol's singular perturbation equation in [DR3]. In each of these examples the singular points have a non-zero linear part: if degenerate, they must be nilpotent. Moreover, only one global blowing-up was needed just for desingularization.

A first step in the direction of the desingularization conjecture is to prove the conjecture for families where any singular point *has a non-zero linear part*. As was said all the bifurcations already studied verify this hypothesis. It includes a singular point of finite codimension whose 1-jet is nilpotent (the codimension 3 was studied in [DRS2] and also the generic "turning points" of singular perturbation equations of any finite codimension). In [PR1] we proved that such families can be reduced to families with an isolated nilpotent point (but perhaps with non-isolated elementary singular points) and we proved a Poincaré-Bendixson theorem in these last families: any limit periodic set must be a *graphic, degenerate or not* (as defined in Chapter 2). Very recently the desingularization conjecture was proved for such families (Panazzolo, thesis [Pa]). This result will be published in a forthcoming paper [PR2]. For instance, the local differential equation at a nilpotent point of finite codimension can be written

$$\dot{x} = y, \quad \dot{y} = (\alpha x^k + \cdots) + y(\beta x^\ell + \cdots) + y^2 Q, \quad \text{with } \alpha, \beta \neq 0 . \qquad (6.63)$$

By blowing-up we do not produce more complicated points than isolated nilpotent one (isolated among the other nilpotent ones). Moreover, it is possible to introduce an index $(k, \ell s)$ at each singular point of the foliated local vector field. This index decreases strictly (in the lexicographic order) at each step of the blow-up. (For nilpotent points of finite codimension (6.63), the index is $(k, \ell, 0)$.)

A general theorem for the desingularization of holomorphic line fields (i.e., holomorphic singular foliation of dimension 1) was proved by Trifonov:

Theorem 34 [Tr] *Let $\pi : E \to \Lambda$ be a holomorphic fibration with a holomorphic line field ℓ tangent to the fibers which are supposed to be 2-dimensional. Then there exist another fibration $\widetilde{\pi} : \widetilde{E} \to \widetilde{\Lambda}$ and holomorphic proper surjective maps $\Phi : \widetilde{E} \to E$ and $\varphi : \widetilde{\Lambda} \to \Lambda$ such that the following diagram is commutative:*

$$
\begin{array}{ccc}
\widetilde{E} & \xrightarrow{\Phi} & E \\
\widetilde{\pi} \big\uparrow & & \big\uparrow \pi \\
\widetilde{\Lambda} & \xrightarrow{\varphi} & \Lambda
\end{array}
\qquad (6.64)
$$

Moreover there exists a line field $\widetilde{\ell}$ on $\widetilde{E}$, whose image $\Phi(\widetilde{\ell})$ is ℓ (for each regular point $\widetilde{p}$ of $\widetilde{\ell}$: $d\Phi(\widetilde{p})(\widetilde{\ell}(\widetilde{p})) = \ell(\Phi(\widetilde{p}))$), and every singular point of $\widetilde{\ell}$ is elementary.

We can translate this theorem into real analytic line fields and also to analytic families of vector fields. However, Trifonov's theorem only looks at the foliation defined by the vector field: if f is an analytic non-trivial function, the vector fields X and fX define the same foliation.

Hence, the transcription of the Theorem 34 for real analytic vector field families is as follows:

Given an analytic vector field family (X_λ) defined on, for instance, $E = S^2 \times \Lambda$, one has an analytic vector field family $(\widetilde{X}_{\widetilde{\lambda}})$ on $\widetilde{E} = \widetilde{S} \times \widetilde{\Lambda}$ and analytic proper surjective maps Φ and φ as above such that $\Phi_*(\widetilde{X}_{\widetilde{\lambda}}) = (X_{\Phi(\widetilde{\lambda})})$. Moreover, for $(\widetilde{p}_0, \widetilde{\lambda}_0) \in \widetilde{E}$, there exists an analytic function $\widetilde{f}$ and a vector field $\widetilde{Y}$ defined in a neighborhood of $\widetilde{p}_0 \in \widetilde{S}$ such that $\widetilde{p}_0$ is a regular or an elementary singular point of $\widetilde{Y}$ and $\widetilde{X}_{\widetilde{\lambda}_0} = \widetilde{f}\widetilde{Y}$.

That is, the zeros of $\widetilde{X}_{\widetilde{\lambda}_0}$ may be non-isolated and, moreover, the factor $\widetilde{f}(\widetilde{p})$ cannot be extended in general to a neighborhood of $(\widetilde{p}_0, \widetilde{\lambda}_0)$ in order to divide the family $\widetilde{X}_\lambda$. As a consequence, non-isolated zeros (in some fiber) are *unavoidable*. In fact, at such points, the family is equivalent to a *singular perturbation equation* by a singular change of time (see [I]). As a simple example we can consider the family $y\,\dfrac{\partial}{\partial y} + \varepsilon\,\dfrac{\partial}{\partial x}$ which has a normally hyperbolic line of zeros at $\varepsilon = 0$.

At a nilpotent point p, the vector field X is written $X = fY$ with $df(p) \neq 0$ and $Y(p) \neq 0$. Thus, nilpotent points are elementary points in the sense of Trifonov's result. The aim in [PR2] is precisely to complete the desingularization of such unfoldings, and in particular to get rid off non-normally-hyperbolic lines of zeros.

In Trifonov's theory we just look at X up to a multiplicative function f, because we are only interested in the foliation. It seems rather clear that the methods used in [Tr] may be used to desingularize Y and f simultaneously, i.e., to obtain a desingularized family where at each point $(\widetilde{p}_0, \widetilde{\lambda}_0)$ $X_{\widetilde{\lambda}_0} = f.Y$ with p_0 a regular or an elementary singular point, and f non-zero or such that $\{f = 0\}$ has normal crossing at $\widetilde{p}_0$, transversal or tangent to the field Y. If this result is true, we will have a finite simple list of possibilities to study. In each case it is possible to write a simple analytic normal form for the germ of X at $p = 0 \in \mathbb{R}^2$:

(a) $X = y^k\,\dfrac{\partial}{\partial x}$ $k \geq 1$ (for $k = 1$: line of nilpotent points).

(b) $X = y^k\,\dfrac{\partial}{\partial y}$ $k \geq 1$ (for $k = 1$: line of normally hyperbolic points).

(c) $X = x^k\,y^\ell\left(x\,\dfrac{\partial}{\partial x} + y\,\dfrac{\partial}{\partial y}\right)$, $k \geq 1$, $\ell \geq 1$ (f has a normal crossing and $Y \neq 0$ is transversal to $\{f = 0\}$).

(d) $X = x^k\,y^\ell\,\dfrac{\partial}{\partial x}$, $k \geq 1$, $\ell \geq 1$ (f has a normal crossing and $Y \neq 0$ is tangent to $\{f = 0\}$).

(e) $X = x^k\,Y$, $k \geq 1$ (Y is a finite codimension semi-hyperbolic point and is tangent to $\{x = 0\}$).

(f) $X = x^k\,y^\ell Y$, $k \geq 1$, $\ell \geq 1$ (Y is as in (e) and tangent to $\{x = 0\}$ and $\{y = 0\}$) (see Figure 6.22).

The point (b), for $k = 1$, is already elementary in the sense of this chapter and (a), for $k = 1$, was treated in [PR1]. The first step in dealing with the other cases would be to obtain a *good normal form theory for the unfoldings*, and next to apply the desingularization method explained in the last section.

Once the desingularization conjecture is proved the second conjecture will remain: prove the finite cyclicity conjecture for elementary limit periodic sets which appear in a foliated local vector field $\mathcal{E}$.

In Chapter 5, we looked at unfoldings of elementary graphics. They correspond to elementary limit periodic sets Γ which belong to the interior of some leaf of $\mathcal{E}$, i.e., such that $\Gamma \cap \Sigma = \emptyset$: in this case $\mathcal{E}$ is equivalent in a neighborhood of Γ to

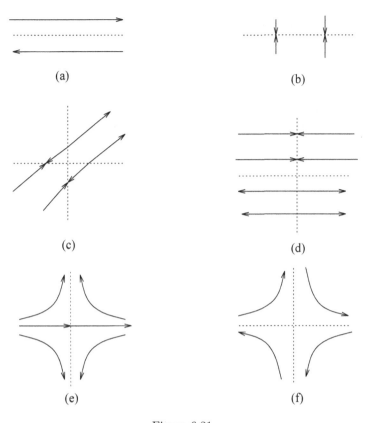

Figure 6.21

a usual unfolding. The study of the cuspidal loop in Section 2 of this chapter gives some idea of the difficulties and the ideas we can use in the general case. A new problem, as we have seen, is to take into account transitions near the elementary points in Γ which are located in the singular set Σ of $\mathcal{E}$.

Bibliography

General references

[ALGM] A. Andronov, E. Leontonich, I. Gordon, A. Maier, *Theory of Bifurcations of Dynamical Systems on a Plane*, Israel Program for Scientific Translations, Jerusalem (1971).

[AA] D. Anosov, V.I. Arnold, *Dynamical Systems I*, Encyclopaedia of Mathematical Sciences, Vol.1, Springer-Verlag (1988).

[A] V.I. Arnold, *Supplementary Chapters to the Theory of Ordinary Differential Equations*, Nauka, Moscow (1978).

[CH] S.-N. Chow, J. Hale, *Methods of Bifurcation Theory*, Springer-Verlag (1982).

[CLW] S.-N. Chow, Chengzhi Li, Duo Wang, *Normal Forms and Bifurcations of Planar Vector Fields*, Cambridge University Press (1994).

[D] F. Dumortier, *Singularities of Vector Fields*, Monografias de Matemática 32, I.M.P.A., Rio de Janeiro (1978).

[GG] M. Golubitsky, V. Guillemin, *Stable Mappings and their Singularities*, GTM 14, Springer-Verlag (1973).

[GH] J. Guckenheimer, P. Holmes, *Non-linear Oscillations, Dynamical Systems, and Bifurcations of Vector Fields*, Appl. Math. Sc. 42, Springer-Verlag (1983).

[MP] W. de Melo, J. Palis, *Geometric Theory of Dynamical Systems*, Springer-Verlag (1982).

[S] D. Schlomiuk, *Bifurcations and Periodic Orbits of Vector Fields*, Nato ASI Series C : Mathematical and Physical Sciences, Vol. 408 (1993).

[So] J. Sotomayor, *Lições de equações diferenciais ordinárias*, Projeto Euclides, CNP$_q$ (1979).

R. Roussarie, *Bifurcations of Planar Vector Fields*
and Hilbert's Sixteenth Problem, Modern Birkhäuser Classics,
DOI: 10.1007/978-3-0348-0718-0, © Springer Basel 1998

References

[AAD] H. Annabi, M.L. Annabi, F. Dumortier, *Continuous dependence on parameters in the Bogdanov-Takens bifurcation*, Pitman Research Notes in Mathematics Series 222, (1992), 1–21.

[Bam2] R. Bamon, *Quadratic vector fields in the plane have a finite number of limit cycles*, Publ. IHES 64 (1986), 111–142.

[Ba] M. Barnsley, *Fractals Everywhere*, Academic Press, San Diego (1988).

[B]   N. Bautin, *On the number of limit cycles which appear by variation of coefficients from an equilibrium position of focus or center type*, Trans. Amer. Math. Soc. 100 (1954), 181–196.

[Bo] R.I. Bogdanov, *Versal deformation of a singularity of a vector field on the plane in the case of zero eigenvalues*, Seminar Petrovski (1976) (Russian), Selecta Math. Soviet. 1 (1981), 389–421 (English).

[Bon] P. Bonckaert, *On the continuous dependence of the smooth change of coodinates in parametrized normal form theorems*, Journal of Differential Equations 106. 1 (1993), 107–120.

[B.M.] M. Brunella, M. Miari, *Topological equivalence of a plane vector field with its principal part defined through Newton polyhedra*, J. Diff. Equ. 85 (1990), 338–366.

[Br] A.D. Brjuno, *Local Methods in Non-linear Differential Equations*, Springer Ser. Soviet Math., Springer-Verlag, Berlin – Heidelberg – New York (1989).

[C]   L.A. Cherkas, *On the stability of singular cycles*, Diff. Equ. 4 (1968), 1012–1017 (Russian).

[Ch] B. Chabat, *Introduction à l'analyse complexe, tome 1*, Mir (ed. française: 1990).

[DeR] Z. Denkowska, R. Roussarie, *A method of desingularization for analytic two-dimensional vector field families*, Bol. Soc. Bras. Mat. Vol. 22, n°1 (1991), 93–126.

[Du1] H. Dulac, *Recherche sur les points singuliers des équations différentielles*, J. Ecole Polytechnique (2), cahier 9 (1904), 1–25.

[Du2] H. Dulac, *Sur les cycles limites*, Bulletin Soc. Math. France 51 (1923), 45–188.

[D1] F. Dumortier, *Singularities of vector fields on the plane*, J. Diff. Equ., 23, n°1 (1977), 53–106.

[D2] F. Dumortier, *Local study of planar vector fields: singularities and their unfoldings*, in: Structures in Dynamics, Finite Dimensional Deterministic Studies (H.W. Broer et al., eds.), Stud. Math. Phys. 2, North-Holland, Amsterdam (1991), 161–241.

[D3] F. Dumortier, *Techniques in the Theory of Local Bifurcations: Blow-Up, Normal Forms, Nilpotent Bifurcations, Singular Perturbations*, Notes written by B. Smits, in [S].

[DER] F. Dumortier, M. El Morsalani, C. Rousseau, *Hilbert's 16th Problem for quadratic vector fields and cyclicity of elementary graphics*, to appear in Non-linearity.

[DR1] F. Dumortier, R. Roussarie, *On the saddle loop bifurcation*, in: Lecture Notes in Math. n° 1455, "Bifurcations of planar vector fields", J.P. Françoise, R. Roussarie, eds. (1990), 44–73.

[DR2] F. Dumortier, R. Roussarie, *Tracking limit cycles escaping from rescaling domains*, in: Proceedings of International Conference on Dynamical Systems and Related Topics, September 1990 – Nagoya Japon – Ed. K. Shiraiva. Adv. Series in Dyn. Syst., World Scientific Vol. 9 (1992), 80–99.

[DR3] F. Dumortier, R. Roussarie, *Duck cycles and centre manifolds*, Memoirs of A.M.S. vol. 121, n°1 (1996), 1–100.

[DRR1] F. Dumortier, R. Roussarie, C. Rousseau, *Hilbert's 16th problem for quadratic vector fields*, Journal of Differential Equations, Vol. 110, n°1 (1994), 86–133.

[DRR2] F. Dumortier, R. Roussarie, C. Rousseau, *Elementary Graphics of cyclicity 1 and 2*, Nonlinearity 7 (1994), 1001–1043.

[DRS1] F. Dumortier, R. Roussarie, J. Sotomayor, *Generic 3-parameter families of vector fields on the plane, unfolding a singularity with nilpotent linear part. The cusp case*, Ergodic Theory Dynamical Systems 7 (1987), 375–413.

[DRS2] F. Dumortier, R. Roussarie, J. Sotomayor, *Generic 3-parameter families of planar vector fields, unfoldings of saddle, focus and elliptic singularities with nilpotent linear parts*, in: Bifurcation of Planar Vector Fields: Nilpotent Singularities and Abelian Integrals (F. Dumortier et al. eds.), Lecture Notes in Math. 1480, Springer-Verlag, Berlin – Heidelberg – New York (1991), 1–164.

[DRS3] F. Dumortier, R. Roussarie, J. Sotomayor, *Generic unfoldings of cuspidal loops*, Preprint n°103, Laboratoire de Topologie, Université de Bourgogne (to appear in Nonlinearity).

[E] J. Ecalle, *Introduction aux fonctions analysables et preuve constructive de la conjecture de Dulac*, Hermann, Paris (1992).

[E1] M. El Morsalani, *Sur la cyclicité de polycycles dégénérées*, Thèse, Université de Bourgogne (1993).

[E2] M. El Morsalani, *Bifurcations de polycycles infinis de champs de vecteurs polynomiaux*, Ann. Fac. Sc. de Toulouse 3 (1994), 387–410.

[E3] M. El Morsalani, *Perturbations of graphics with semi-hyperbolic singularities*, Preprint n°35, Laboratoire de Topologie, Université de Bourgogne (1995), to appear in Bulletin de Sciences Mathématiques.

[EM] M. El Morsalani, A. Mourtada, *Degenerate and non-trivial hyperbolic 2-polycycles: appearance of 2 independent Ecalle-Roussarie compensators and Khovanskii's theory*, Nonlinearity 7 (1994), 1593–1604.

[EMR] M. El Morsalani, A. Mourtada, R. Roussarie, *Quasi-regularity property for unfoldings of hyperbolic polycycles*, Astérisque: "Complex analytic methods in dynamical system". C. Camacho and al ed., n° 222 (1994), 303–326.

[F] J.P. Françoise, *Successive derivatives of a first return map, application to the study of quadratic vector fields*, Preprint, Paris 6 (1993).

[FP] J.P. Françoise, C.C. Pugh, *Keeping track of limit cycles*, J. Differential Equations 65 (1986), 139–157.

[G] A.M. Gabrielov, *Projections of semi-analytic sets*, Functional Anal. Appl. 2 (1968), 282–291.

[H] M. Hervé, *Several Complex Variables*, Oxford University Press (1963).

[HI] E. Horozov, I.D. Ilev, *On saddle-loop bifurcations of limit cycles in perturbations of quadratic Hamiltonian systems*, J. Diff. Equ., vol.113, no 1, (1994), 84–105.

[I1] Yu. Il'Yashenko, *An example of equation $dw/dz = P_n(z,w)/Q_n(z,w)$ having a countable number of limit cycles and an arbitrarily large genre after Petrovskii-Landis*, USSR Math. Sbor., 80, 3, (1969), 388–404.

[I2] Yu. Il'Yasenko, *Limit cycles of polynomial vector fields with nondegenerate singular points on the real plane*, Funk. Anal. Ego. Pri., 18, 3,(1984), 32–34. (Func. Ana. and Appl., 18,3 (1985), 199–209).

[I3] Yu. Il'Yashenko, *Finiteness theorems for limit cycles*, Amer. Math. Soc., Providence, RI (1991).

[I4] Yu. Il'Yashenko, *Local Dynamics and Nonlocal Bifurcations*, notes by A.M. Arkhipov and A.I. Shilov in [S].

[IY1] Yu. Il'Yashenko, S. Yakovenko, *Finitely-smooth normal forms for local families of diffeomorphisms and vector fields*, Russian Math. Surveys 46 n° 1 (1991), 1–43.

[IY2] Yu. Il'Yashenko, S. Yakovenko, *Concerning the Hilbert Sixteenth Problem*, Advances in Mathematical Sciences-23, Amer. Math. Soc. Translations, Series 2, vol. 165, AMS Publ. (1995), 1–20.

[IY3] Y. Il'Yashenko, S. Yakovenko, *Finite cyclicity of elementary polycycles in generic families*, Concerning the Hilbert Sixteenth Problem, Advances in Mathematical Sciences-23, Amer. Math. Soc. Translations, Series 2, vol. 165, AMS Publ. (1995), 21–95.

[JM] A. Jebrane, A. Mourtada, *Cyclicité finie des lacets doubles non triviaux*, Nonlinearity 7, n° 22 (1994), 1349–1365.

[JZ] A. Jebrane, H. Zoladek, *A non symmetric perturbation of symmetric hamiltonian vector field*, Advan. and Appl. Math. n° 15 (1994), 1–12.

[J1] P. Joyal, *Un théorème de préparation pour les fonctions a développement tchébychévien*, Erg. Theory Dynamical Sys. (to appear).

[J2] P. Joyal, *The generalized homoclinic bifurcation*, J. Diff. Equ. 107 (1994), 1–45.

[J3] P. Joyal, *The cusp of order N*, J. Diff. Equ. 88 (1990), 1–14.

[K]   A. Khovanskii, *Fewnomials*, Amer. Math. Soc., Providence, RI (1991).

[KS] A. Kotova, V. Stanzo, *On few-parametric generic families of vector fields on the two-dimensional sphere*, Concerning the Hilbert Sixteenth Problem, Advances in Mathematical Sciences-23, Amer. Math. Soc. Translations, Series 2, vol. 165, AMS Publ. (1995), 155–201.

[M]   B. Malgrange, *Ideals of Differentiable Functions*, Oxford University Press (1966).

[Mar1] P. Mardesić, *The number of limit cycles of polynomial deformations of a Hamiltonian vector field*, Ergodic Theory Dynamical Systems 10 (1990), 523–529.

[Mar2] P. Mardesić, *An explicit bound for the multiplicity of zeros of generic Abelian integrals*, Nonlinearity 4 (1991), 845–852.

[Mar3] P. Mardesić, *Le déploiement versel du cusp d'ordre n*, Thèse, Université de Bourgogne (1992).

[Ma] J. Martinet, *Singularités des Fonctions et applications différentiables*, P.U.C. (1977).

[M1] A. Mourtada, *Cyclicité finie des polycycles hyperboliques des champs de vecteurs du plan: mise sous forme normale*, in: Bifurcations of Planar Vector Fields (J.P. Françoise and R. Roussarie, eds.), Lecture Notes in Math. 1455, Springer-Verlag, Berlin – Heidelberg – New York (1990), 272–314.

[M2] A. Mourtada, *Cyclicité finie des polycycles hyperboliques de champs de vecteurs du plan: algorithme de finitude*, Ann. Inst. Fourier (Grenoble) 41 (1991), 719–753.

[M3] A. Mourtada, *Degenerate and non-trivial hyperbolic polycycles with two vertices*, Preprint, Laboratoire de Topologie, Université de Bourgogne (1991).

[M4] A. Mourtada, *Analytic Unfolding of Irrational and Trivial 2-polycycle*, Preprint, Laboratoire de Topologie n° 16, (1992).

[Mo] R. Moussu, *Le problème de la finitude du nombre de cycles limites*, Astérisque 145–146 volume 1985/1986, exposés 651–668, (1987), 89–99.

[MoR] R. Moussu, C. Roche, *Théorème de Khovanskii et problème de Dulac*, Invent. Math., 105 (1991), 431–441.

[N]   R.Narashiman, *Introduction to the Theory of Analytic Spaces*, Lect. Notes in Math. 25 (1966).

[Pa] D. Panazzolo, *Desingularization of Nilpotent Singularities in Analytic Families*, Thèse, Université de Bourgogne (1997).

[PR1] D. Panazzolo, R. Roussarie, *A Poincaré-Bendixson theorem for analytic families of vector fields*, Bol. Soc. Br. Mat., Vol.26, N. 1 (1995), 85–116.

[PR2] D. Panazzolo, R. Roussarie, *Desingularization of analytic families of planar vector fields without singular point of vanishing linear part*, (in preparation).

[P]   M. Peixoto, *Structural stability on two-dimensional manifolds*, Topology 1 (1962).

[Per] L.Perko, *Differential Equations and Dynamical Systems*, Appl. Math. 7, Springer-Verlag (1993)

[Pe] G.S. Petrov, *Elliptic integrals and their non-oscillation*, Functional Anal. Appl. 20 (1985), 37–40.

[Po] J.C. Poggiale, *Applications des variétés invariantes à la modélisation de l'hétérogénéité en dynamique des populations*, Thèse, Université de Bourgogne (1994).

[Poi] H. Poincaré, *Les méthodes nouvelles de la mécanique céleste* , 3 Vols. Gauthier Villars (1899).

[RE] R. Roussarie, *Techniques in theory of local bifurcations: Cyclicity and desingularization*, 347–382 in [S].

[R1] R. Roussarie, *Weak and continuous equivalences for families of line diffeomorphisms in dynamical systems and bifurcation theory*, Camacho, Pacifico ed., Longman, Scientific and Technical, Pitman Research Notes in Math. Series 160 (1987), 377–385.

[R2] R. Roussarie, *Déformations génériques des cusps*, Astérisque n° 150/151: Singularités d'Equations Différentielles, Dijon 1985, (1987), 151–184.

[R3] R. Roussarie, *A note on finite cyclicity property and Hilbert's 16th problem*, Lecture Notes in: Math. 1331, "Dynamical Systems-Valparaiso 1986", R. Bamon, R. Labarca, J. Palis, eds. (1988).

[R4] R. Roussarie, *On the number of limit cycles which appear by perturbation of separatrix loop of planar vector fields*, Bol. Soc. Bras. Mat., 17, (1986), no. 2, 67–101.

[R5] R. Roussarie, *Cyclicité finie des lacets et des points cuspidaux*, Nonlinearity, fasc. 2, (1989), 73–117.

[R6] R. Roussarie, *Desingularization of unfoldings of cuspidal loops*, in: "Geometry and analysis in nonlinear dynamics" (with H.W. Broer, F. Takens) Eds. Pitman Research Notes in Math. Series, n° 222, Longman Scientific and Technical (1992), 41–55.

[R7] R. Roussarie, *Smoothness properties for bifurcation diagrams*, Publicacions Matematiques, Vol. 41 (1996), 243–268.

[RW] R. Roussarie, F. Wagener, *A study of the Takens-Bogdanov bifurcation*, Resenhas do IME-USP, vol. 2, n°1 (1995), 1–25.

[Ro] C. Rousseau, *Bifurcation methods in polynomial systems*, 383–428 in [S].

[RoZ] C. Rousseau, H. Zoladek, *Zeroes of complete elliptic integrals for 1:2 resonance*, J.Diff.Eq 94 (1991), 41–54.

[S1] D. Schlomiuk, J. Guckenheimer, R. Rand, *Integrability of plane quadratic vector fields*, Exposition. Math. 8 (1990), 3–25.

[Sm] S. Smale, *Structurally stable systems are non dense*, Amer. Jour. Math., 88, (1966), 491–496.

[So1] J. Sotomayor, *Generic one-parameter families of vector fields on two-dimensional manifolds*, Publ. Math. I.H.E.S. 43 (1974), 5–46.

[So2] J. Sotomayor, *On stable planar polynomial vector fields*, Rev. Mat. Ibero-americana, 2 (1985), 15–23.

[T1] F. Takens, *Unfoldings of certain singularities of vector fields. Generalized Hopf bifurcations*, J. Diff. Equ. 14 (1973), 476–493.

[T2] F. Takens, *Forced oscillations and bifurcations*, in: Applications of Global Analysis I, Comm. Math. Inst. University of Utrecht 3 (1974), 1–59.

[T3] F. Takens, *Singularities of vector fields*, Publ. IHES 43 (1974), 48–100.

[Tr] S. Trifonov, *Desingularization in families of analytic differential equations*, Ann. Math. Soc. Trans. (2), Vol. 165 (1995), 97–129.

[Y] S. Yakovenko, *A geometric proof of Bautin theorem*, Concerning the Hilbert Sixteenth Problem, Advances in Mathematical Sciences-23, Amer. Math. Soc. Translations, series 2, vol. 165, AMS Publ., Providence, RI, (1995), 203–219.

[Ye] Y. Ye, *Theory of Limit Cycles*, Amer. Math. Soc., Providence, RI (1986). Monographs, Vol. 66, Trans. Math.

[Z] H. Zoladek, *On cyclicity of triangles in quadratic systems*, J. Diff. Equ. 122, n°1 (1995), 137–159.

Index

Abelian integrals, 15, 82
absolute cyclicity, 23
adapted to 0, 130
admissible
 − rational functions, 114
 − trivialization, 186
algebra of admissible functions, 113
algebraically isolated singularity, 37
analytic
 − compact family, 24
 − vector field, 4
asymptotic expansion, 33

Bautin Ideal $\mathcal{I}$, 51, 60, 63
Bendixson compactification, 5
bifurcation set, 7
blown-up foliated local
 vector field, 186
Bogdanov-Takens bifurcation, 10

$(\mathcal{C}^0$-fibre, $\mathcal{C}^s)$-equivalent, 7
$(\mathcal{C}^0, \mathcal{C}^s)$-equivalent, 8
$\mathcal{C}^0$-equivalent, 8
$\mathcal{C}^\infty$-equivalent, 57
$(\mathcal{C}_1)$, 141
$(\mathcal{C}_2)$, 141
$(\mathcal{C}_3)$, 141
$\mathcal{C}(E)$, 186
$\mathcal{C}(M)$, 18
$\mathcal{C}^0$-fibre equivalent, 8
$\mathcal{C}^\infty$ conjugacy, 57
$\mathcal{C}^s$-structurally stable, 4
$\mathcal{C}ycl\ (X_\lambda, \Gamma)$, 22

center-type, 59
 − point, 52
characteristic orbits, 38
$\chi^\infty(S)$, 55
$\chi^r(S; P)$, 7
class
 − I, 107
 − I_λ^k for $(x - \eta(\lambda))$, 144
compact family, 24
compensator, 141
critical locus, 186
cusp singularity, 11
cuspidal loops, 155
cyclicity, 7, 22
 − property, 22

$D(x)$, 44
DC^{in}, 157
DC^{out}, 157
$D_\lambda(x)$, 105
$\mathcal{D}$, 45, 108
$\mathcal{D}_k$, 108
$\widehat{D}$, 105
$\widehat{D}(x)$, 44
degenerate graphics, 22
$\delta(u, \lambda)$, 52
$\delta_\lambda(u)$, 52
desingularization, 33
 − conjecture, 187
 − operations, 181
 − step, 187
 − theorem, 37
desingularized vector field, 154

201